A FIRST COURSE OF
HOMOLOGICAL ALGEBRA

A FIRST COURSE OF
HOMOLOGICAL ALGEBRA

D. G. NORTHCOTT, F.R.S.

Town Trust Professor of Pure Mathematics
University of Sheffield

CAMBRIDGE UNIVERSITY PRESS

CAMBRIDGE UNIVERSITY PRESS
Cambridge, New York, Melbourne, Madrid, Cape Town, Singapore, São Paulo, Delhi

Cambridge University Press
The Edinburgh Building, Cambridge CB2 8RU, UK

Published in the United States of America by Cambridge University Press, New York

www.cambridge.org
Information on this title: www.cambridge.org/9780521201964

First published 1973
Re-issued in this digitally printed version 2008

A catalogue record for this publication is available from the British Library

Library of Congress Catalogue Card Number: 72-97873

ISBN 978-0-521-20196-4 hardback
ISBN 978-0-521-29976-3 paperback

CONTENTS

PREFACE

The main part of this book is an expanded version of lectures which I gave at Sheffield University during the session 1971-2. These lectures were intended to provide a first course of Homological Algebra, assuming only a knowledge of the most elementary parts of the theory of modules. The amount of time available was very limited and ruled out any approach which required the elaborate machinery or great generality that is sometimes associated with the subject. The alternative, it seemed to me, was to build the course round a number of topics which I hoped my audience would find interesting, and create the necessary tools by *ad hoc* constructions. Fortunately it proved rather easy to find topics where the techniques needed to treat one of them could also be used on the others. In the event, the first five chapters were fully covered in the course. The last chapter was added later and it differs from those that precede it by including some material which, so far as I am aware, has not previously appeared in print. This material has to do with what are here called *semi-commutative local algebras*. It is hoped that it may be of some interest to the specialist as well as to the beginner.

Reference has already been made to one way in which the amount of available time influenced the structure of the course. It had, indeed, a second effect. In order to speed up the presentation, some easily proved results and parts of some demonstrations were left as exercises. Other exercises were included in order to expand the main themes. What actually happened was that two members of the class, Mr A. S. McKerrow and Mr P. M. Scott, were good-natured enough to do all the exercises and, in addition, they provided the other participants with copies of their solutions. These solutions, edited so as to remove differences of style, are reproduced here. However the reader will find that his grasp of the subject is much improved if he works out a fair proportion of the problems for himself, rather than merely checks through the details of the arguments provided. The more difficult exercises have been marked with an asterisk.

I am much indebted to other mathematicians who have written on similar or related topics, and the list of references at the end shows the books and papers that I have consulted recently. It is a pleasure to acknowledge the help and benefit that I have derived from these and other sources. I have not attempted to compile a comprehensive bibliography. Naturally the degree of my indebtedness varies from one author to another. I have, for example, made much use of I. Kaplansky's treatment of homological dimension. Also I am very conscious of the influence which the writings of H. Bass and R. G. Swan have had on this account.

As on other occasions, I have been very fortunate in the help that has been given to me. Once again my secretary, Mrs E. Benson, has converted pages of untidy manuscript into an orderly form where the idea that they might turn into a book no longer seemed unreasonable. Besides this Mr A. S. McKerrow checked much of the first draft to see that it was technically correct. Their assistance has been extremely valuable and I am most grateful to them both.

D. G. NORTHCOTT

Sheffield
October 1972

NOTES FOR THE READER

This opportunity is taken to summarize what the reader is assumed to know already, and to draw his attention to any conventions or terminology which may differ slightly from those to which he has been accustomed.

All the main topics in this book have to do with *rings* and *modules*. First a word about rings. Unless otherwise stated, these need not be commutative, but every one is required to have an identity element. (Usually the identity element does not have to be different from the zero element.) When we speak of a homomorphism of one ring into another, it is to be understood that the identity element of the former is mapped into that of the latter. In particular, if Γ is a *subring* of a ring Λ, that is if the inclusion mapping $\Gamma \to \Lambda$ is a ring-homomorphism, then our convention ensures that Γ and Λ must have the same identity element. An important subring of Λ is its *centre*. This, of course, is composed of all elements γ with the property that $\lambda\gamma = \gamma\lambda$ for every λ in Λ.

Let Λ be a ring. In any reference to a Λ-*module* it is always intended that multiplication of an element of the module by the identity 1_Λ, of Λ, shall leave the element of the module unchanged. In other words, we only consider *unitary* modules. Note that there are two *types* of Λ-module, namely *left* Λ-modules and *right* Λ-modules.† The system formed by all left resp. right Λ-modules (and the homomorphisms between them) is referred to as the *category* of left resp. right Λ-modules and is denoted by $\mathscr{C}_\Lambda^L$ resp. $\mathscr{C}_\Lambda^R$. Though use is made of the language of Category Theory it is not at all necessary that the reader should have previously met the definition of an abstract category. To illustrate the language let us observe that a module over the ring Z of integers is just the same as an (additively written) abelian group. Further if A and B are two such objects, then a mapping $f: A \to B$ is a homomorphism of Z-modules if and only if it is a group-homomorphism. A convenient way in which to describe all this is to say that *the category of Z-modules can be identified with the category of (additively written) abelian groups*.

Although we assume no general knowledge of Category Theory it is

---

† If the ring is *commutative* we do not need to make this distinction.

supposed that the reader is familiar with the elementary theory of modules and, on this basis, certain terms are used without explanation. The following are typical examples: *submodule, factor module*; *image*; *kernel* and *cokernel* (of a homomorphism); *exact sequence, commutative diagram*; *direct sum* and *direct product*. In addition we take as known the standard *isomorphism theorems* and presuppose some elementary knowledge of transfinite methods based on well-ordering and *Zorn's Lemma*. A leisurely account of these matters will be found in (20) in the list of references, should the reader wish to supplement his knowledge.

Let $f: A \to B$ be a homomorphism of Λ-modules. If, in addition, f is an injective mapping, then, of course, it is customary to say that f is a *monomorphism*. We shall also say that f is *monic* whenever we wish to describe a situation of this kind. This is done solely to expand a limited vocabulary which otherwise could lead to tedious repetition. For the same reason, if the homomorphism f is a surjective mapping, then we shall say either that f is an *epimorphism* or that it is *epic* depending on which alternative description happens to be the more convenient.

Our next remarks concern notation in relation to sets and modules. Thus if A is a set, then i_A always denotes the *identity mapping* of A. Now suppose that X and Y are sets. If X is a subset of Y and we wish to indicate this, then we shall write $X \subseteq Y$. However, if X is a *proper subset* of Y, that is if $X \subseteq Y$ but $X \neq Y$, then $X \subset Y$ will be used to convey this information.

Turning now to modules, let Λ be a ring and $\{A_i\}_{i \in I}$ a family of Λ-modules. The family will have both a direct sum and a direct product. The former of these will be denoted by $\bigoplus_{i \in I} A_i$ and the latter by $\prod_{i \in I} A_i$. However when we have to do with a finite family

$$\{A_1, A_2, ..., A_n\},$$

then we use $A_1 \oplus A_2 \oplus ... \oplus A_n$ and $A_1 \times A_2 \times ... \times A_n$ as alternatives to $\bigoplus_{i=1}^{n} A_i$ and $\prod_{i=1}^{n} A_i$ respectively. Again if A is a Λ-module, then $\bigoplus_{i \in I} A$ or $\bigoplus_{I} A$ will denote a direct sum in which all the summands are equal to A and there is one of them for each member of I. Likewise $\prod_{i \in I} A$ or $\prod_{I} A$ will denote a direct product in which each factor is A and there is one factor for each element of the set I.

It is hoped that enough has now been said to prepare the reader. Note that the numbering of theorems, lemmas and so on is begun afresh in each chapter. If a reference is made to a result and no chapter or section is specified, then the result in question is to be found in the chapter being read. In all other cases the extra information needed for identification is provided.

1

THE LANGUAGE OF FUNCTORS

1.1 Notation

Λ, Γ, Δ will denote rings with identity elements. They need not be commutative. Z will denote the ring of integers. The category of left (resp. right) Λ-modules will be denoted by $\mathscr{C}_\Lambda^L$ (resp. $\mathscr{C}_\Lambda^R$). Sometimes it is immaterial whether we work exclusively with left Λ-modules or exclusively with right Λ-modules. In such a case $\mathscr{C}_\Lambda$ will denote the category in question. When Λ is commutative, we make no distinction between $\mathscr{C}_\Lambda^L$ and $\mathscr{C}_\Lambda^R$. Also we normally identify the category of additively written abelian groups with the category of Z-modules. Finally i_A is used to denote the identity map of A.

1.2 Bimodules

Suppose that A is both a Λ-module and a Γ-module, the additive structure being the same in both cases. Let us suppose that multiplication (of an element of A) by an element of Λ always commutes with multiplication by an element of Γ. We then say that A is a (Λ, Γ)-*bimodule*. If, for example, Λ operates on the left and Γ on the right, we may indicate this by writing ${}_\Lambda A_\Gamma$. If A and A' are both (Λ, Γ)-bimodules of the same type, then a mapping $f: A \to A'$ which is simultaneously Λ-linear and Γ-linear is called a *bihomomorphism*.

Example 1. Every Λ-module is a (Λ, Z)-bimodule.

Example 2. If Γ is the centre of Λ, then every Λ-module is a (Λ, Γ)-bimodule.

Example 3. Λ itself is a (Λ, Λ)-bimodule with one Λ acting on the right and the other on the left. This is by virtue of the associative law of multiplication.

1.3 Covariant functors

Suppose that with each module A in $\mathscr{C}_\Lambda$ there is associated a module $F(A)$ in $\mathscr{C}_\Delta$ and that to each Λ-homomorphism $f: A \to A'$ there corresponds a Δ-homomorphism $F(f): F(A) \to F(A')$. Suppose further that

(1) $F(i_A) = i_{F(A)}$ for all A in $\mathscr{C}_\Lambda$;

(2) $F(gf) = F(g) F(f)$ whenever $f: A \to A'$ and $g: A' \to A''$ in $\mathscr{C}_\Lambda$.

In these circumstances we say we have a *covariant functor* $F: \mathscr{C}_\Lambda \to \mathscr{C}_\Delta$ from Λ-modules to Δ-modules. Simple commutative diagrams (of Λ-modules and Λ-homomorphisms) such as

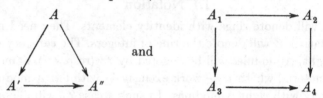

remain commutative when a covariant functor is applied. Also if $f: A \to A'$ is an isomorphism and $g: A' \to A$ is its inverse, then, for a covariant functor F, $F(f): F(A) \to F(A')$ is an isomorphism and $F(g): F(A') \to F(A)$ is its inverse. This is because gf and fg are identity maps.

For the remainder of section (1.3), $F: \mathscr{C}_\Lambda \to \mathscr{C}_\Delta$ will denote a co-variant functor.

Definition. *F is said to be 'additive' if whenever $f_1: A \to A'$ and $f_2: A \to A'$ are Λ-homomorphisms, sharing a common domain A and a common codomain A', we have $F(f_1 + f_2) = F(f_1) + F(f_2)$.*

Note. The Λ-homomorphisms of A into A' form an abelian group. This is denoted by $\mathrm{Hom}_\Lambda(A, A')$. Addition in $\mathrm{Hom}_\Lambda(A, A')$ is defined by $(f_1 + f_2)(a) = f_1(a) + f_2(a)$.

If F is additive, then it carries null homomorphisms and null modules into null homomorphisms and null modules.

In the classical theory of modules, *finite* direct sums and *finite* direct products are indistinguishable. Here this is recognized by introducing the notion of a *biproduct*.

Let $A_1, A_2, ..., A_n$ and A be Λ-modules and suppose we are given homomorphisms $\sigma_i: A_i \to A (1 \leqslant i \leqslant n)$ and $\pi_i: A \to A_i (1 \leqslant i \leqslant n)$. The complete system is called a representation of A as a *biproduct* of $A_1, A_2, ..., A_n$ if

(a) $\pi_j \sigma_i = \delta_{ji}$, i.e. $\pi_j \sigma_i$ is a null resp. identity homomorphism if $i \neq j$ resp. $i = j$;

(b) $\Sigma \sigma_i \pi_i =$ identity.

In these circumstances we write variously

$$A = A_1 \oplus A_2 \oplus \dots \oplus A_n \text{ (direct sum notation)},$$
$$A = A_1 \times A_2 \times \dots \times A_n \text{ (direct product notation)},$$
$$A = A_1 * A_2 * \dots * A_n \text{ (biproduct notation)},$$

and, more explicitly,

$$[\sigma_1, \dots, \sigma_n; A; \pi_1, \dots, \pi_n] = A_1 * A_2 * \dots * A_n.$$

We call $\sigma_i : A_i \to A$ the *canonical injection* (it is necessarily a monomorphism) and $\pi_i : A \to A_i$ the *canonical projection* (it is necessarily an epimorphism).

Exercise 1.† *Let* $[\sigma_1, \dots, \sigma_n; A; \pi_1, \dots, \pi_n] = A_1 * A_2 * \dots * A_n$ *in* $\mathscr{C}_\Lambda$. *Show that if Λ-homomorphisms $f_i : A_i \to B$ $(1 \leqslant i \leqslant n)$ are given, then there exists a unique homomorphism $f : A \to B$ such that $f\sigma_i = f_i$ for $1 \leqslant i \leqslant n$. Show also that if $g_i : B \to A_i$ $(1 \leqslant i \leqslant n)$ are prescribed Λ-homomorphisms, then there exists a unique homomorphism $g : B \to A$ such that $\pi_i g = g_i$ for $1 \leqslant i \leqslant n$.*

Exercise 2. *Let* $[\sigma_1, \dots, \sigma_n; A; \pi_1, \dots, \pi_n] = A_1 * A_2 * \dots * A_n$ *in* $\mathscr{C}_\Lambda$. *Show that the homomorphism $A_1 \oplus A_2 \oplus \dots \oplus A_n \to A$ induced by the σ_i and the homomorphism $A \to A_1 \times A_2 \times \dots \times A_n$ induced by the π_i are both of them isomorphisms.*

Observe that if $A_1, A_2, \dots, A_n$ are given, then we can always find $A, \sigma_1, \sigma_2, \dots, \sigma_n$ and $\pi_1, \pi_2, \dots, \pi_n$ so that

$$[\sigma_1, \dots, \sigma_n; A; \pi_1, \dots, \pi_n] = A_1 * A_2 * \dots * A_n.$$

Theorem 1. *Let $F : \mathscr{C}_\Lambda \to \mathscr{C}_\Delta$ be an additive covariant functor and let* $[\sigma_1, \dots, \sigma_n; A; \pi_1, \dots, \pi_n] = A_1 * A_2 * \dots * A_n$ *in* $\mathscr{C}_\Lambda$. *Then*
$$[F(\sigma_1), \dots, F(\sigma_n); F(A); F(\pi_1), \dots, F(\pi_n)] = F(A_1) * F(A_2) * \dots * F(A_n)$$
in $\mathscr{C}_\Delta$.

Proof. Apply F to the relations $\pi_j \sigma_i = \delta_{ji}$ and $\Sigma \sigma_i \pi_i = $ identity.

We shall now show that this property characterizes additive covariant functors.

Theorem 2. *Let $F : \mathscr{C}_\Lambda \to \mathscr{C}_\Delta$ be a covariant functor and suppose that whenever* $[\sigma_1, \sigma_2; A; \pi_1, \pi_2] = A_1 * A_2$ *in* $\mathscr{C}_\Lambda$, *then*

$$[F(\sigma_1), F(\sigma_2); F(A); F(\pi_1), F(\pi_2)] = F(A_1) * F(A_2) \text{ in } \mathscr{C}_\Lambda.$$

In these circumstances F is additive.

† Solutions to the Exercises will be found at the end of the chapter.

Proof. Let $f_1, f_2: A \to B$ be homomorphisms. Further, let

$$[\sigma_1, \sigma_2; C; \pi_1, \pi_2] = A * A.$$

Then $[F(\sigma_1), F(\sigma_2); F(C); F(\pi_1), F(\pi_2)] = F(A) * F(A)$ and therefore $i_{F(C)} = F(\sigma_1) F(\pi_1) + F(\sigma_2) F(\pi_2)$. Define $d: A \to C$ by $d = \sigma_1 + \sigma_2$. Then $\pi_1 d = \pi_1(\sigma_1 + \sigma_2) = i_A$ from which we obtain

$$F(\pi_1) F(d) = F(\pi_1 d) = F(i_A) = i_{F(A)}.$$

Similarly $F(\pi_2) F(d) = i_{F(A)}$. Now

$$F(d) = i_{F(C)} F(d) = (F(\sigma_1) F(\pi_1) + F(\sigma_2) F(\pi_2)) F(d).$$

Hence

$$F(d) = F(\sigma_1) F(\pi_1) F(d) + F(\sigma_2) F(\pi_2) F(d)$$
$$= F(\sigma_1) i_{F(A)} + F(\sigma_2) i_{F(A)} = F(\sigma_1) + F(\sigma_2).$$

Define $g: C \to B$ by $g = f_1 \pi_1 + f_2 \pi_2$. Then

$$g\sigma_1 = (f_1 \pi_1 + f_2 \pi_2) \sigma_1 = f_1 \pi_1 \sigma_1 + f_2 \pi_2 \sigma_1 = f_1.$$

Similarly $g\sigma_2 = f_2$. Furthermore $gd = (f_1 \pi_1 + f_2 \pi_2)(\sigma_1 + \sigma_2) = f_1 + f_2$. Accordingly $F(f_1 + f_2) = F(gd) = F(g) F(d) = F(g)(F(\sigma_1) + F(\sigma_2))$. Thus

$$F(f_1 + f_2) = F(g) F(\sigma_1) + F(g) F(\sigma_2)$$
$$= F(g\sigma_1) + F(g\sigma_2)$$
$$= F(f_1) + F(f_2).$$

Hence f is additive.

Theorem 3. *Suppose that* $[\sigma_1, \sigma_2; A; \pi_1, \pi_2) = A_1 * A_2$ *in* $\mathscr{C}_\Lambda$. *Then the sequences*

$$0 \to A_1 \overset{\sigma_1}{\to} A \overset{\pi_2}{\to} A_2 \to 0 \qquad (1.3.1)$$

and

$$0 \to A_2 \overset{\sigma_2}{\to} A \overset{\pi_1}{\to} A_1 \to 0 \qquad (1.3.2)$$

are exact.

Proof. We need only consider (1.3.1) and for this it suffices to show $\mathrm{Ker}\, \pi_2 \subseteq \mathrm{Im}\, \sigma_1$. Let $\alpha \in \mathrm{Ker}\, \pi_2$. Then

$$\alpha = \sigma_1 \pi_1(\alpha) + \sigma_2 \pi_2(\alpha) = \sigma_1 \pi_1(\alpha) \in \mathrm{Im}\, \sigma_1.$$

Lemma 1. *Suppose that* $A_1 \overset{\sigma_1}{\to} A$ *and* $A \overset{\pi_1}{\to} A_1$ *are* Λ-*homomorphisms such that* $\pi_1 \sigma_1 = identity$. *Then* $A = \mathrm{Im}\, \sigma_1 \oplus \mathrm{Ker}\, \pi_1$.

Proof. Let $a \in A$. Then $\pi_1(a - \sigma_1 \pi_1(a)) = 0$ and therefore

$$a = \sigma_1 \pi_1(a) + (a - \sigma_1 \pi_1(a)) \in \mathrm{Im}\, \sigma_1 + \mathrm{Ker}\, \pi_1.$$

Now assume that $\alpha \in \operatorname{Im} \sigma_1 \cap \operatorname{Ker} \pi_1$, say $\alpha = \sigma_1(a_1)$ with $a_1 \in A_1$. Then

$$a_1 = \pi_1 \sigma_1(a_1) = \pi_1(\alpha) = 0$$

whence $\alpha = 0$. This shows that $A = \operatorname{Im} \sigma_1 \oplus \operatorname{Ker} \pi_1$.

Theorem 4. *Let* $0 \to A_1 \overset{\sigma_1}{\to} A \overset{\pi_2}{\to} A_2 \to 0$ *be an exact sequence in* $\mathscr{C}_\Lambda$. *Then the following statements are equivalent*:

(1) $\operatorname{Im} \sigma_1 (= \operatorname{Ker} \pi_2)$ *is a direct summand of* A;
(2) *there exists a* Λ-*homomorphism* $\pi_1 : A \to A_1$ *such that* $\pi_1 \sigma_1 = identity$;
(3) *there exists a* Λ-*homomorphism* $\sigma_2 : A_2 \to A$ *such that* $\pi_2 \sigma_2 = identity$;
(4) *there exist* Λ-*homomorphisms* $\sigma_2 : A_2 \to A$ *and* $\pi_1 : A \to A_1$ *such that* $[\sigma_1, \sigma_2 ; A ; \pi_1, \pi_2] = A_1 * A_2$.

Proof. By the definitions and Lemma 1,

$$(4) \Rightarrow (2) \Rightarrow (1) \quad \text{and} \quad (4) \Rightarrow (3) \Rightarrow (1).$$

Assume (1), say $A = \operatorname{Im} \sigma_1 \oplus B$ for some submodule B of A. Now σ_1 induces an isomorphism $A_1 \overset{\sim}{\to} \operatorname{Im} \sigma_1$. Let $u : \operatorname{Im} \sigma_1 \overset{\sim}{\to} A_1$ be its inverse. Next π_2 induces an isomorphism $B \overset{\sim}{\to} A_2$. Let $v : A_2 \overset{\sim}{\to} B$ be its inverse. Put $\pi_1 = up$ and $\sigma_2 = jv$, where $p : A \to \operatorname{Im} \sigma_1$ is the projection associated with the relation $A = \operatorname{Im} \sigma_1 \oplus B$ and $j : B \to A$ is an inclusion mapping. Then $\pi_1 \sigma_1 = $ identity, $\pi_1 \sigma_2 = 0$, $\pi_2 \sigma_1 = 0$, $\pi_2 \sigma_2 = $ identity. Finally if $a \in A$, then $\sigma_1 \pi_1(a)$ is the projection of a on $\operatorname{Im} \sigma_1$ and $\sigma_2 \pi_2(a)$ is the projection of a on B. Thus

$$\sigma_1 \pi_1(a) + \sigma_2 \pi_2(a) = a \quad \text{or} \quad \sigma_1 \pi_1 + \sigma_2 \pi_2 = \text{identity}.$$

Accordingly (1) implies (4).

Definition. *Let* $0 \to A_1 \overset{\sigma_1}{\to} A \overset{\pi_2}{\to} A_2 \to 0$ *be an exact sequence in* $\mathscr{C}_\Lambda$. *If the four equivalent conditions of Theorem 4 hold, then it is called a 'split exact sequence'.*

We now see, in view of Theorem 3, that if $[\sigma_1, \sigma_2 ; A ; \pi_1, \pi_2] = A_1 * A_2$, then

$$0 \to A_1 \overset{\sigma_1}{\to} A \overset{\pi_2}{\to} A_2 \to 0$$

and

$$0 \to A_2 \overset{\sigma_2}{\to} A \overset{\pi_1}{\to} A_1 \to 0$$

are split exact sequences. On the other hand, if $0 \to B \to A \to C \to 0$ is a split exact sequence, then we have an isomorphism $A \approx B \oplus C$.

Theorem 5. *Let* $0 \to A_1 \overset{\sigma_1}{\to} A \overset{\pi_2}{\to} A_2 \to 0$ *be a split exact sequence in* $\mathscr{C}_\Lambda$ *and* $F: \mathscr{C}_\Lambda \to \mathscr{C}_\Delta$ *an additive covariant functor. Then*

$$0 \to F(A_1) \overset{F(\sigma_1)}{\longrightarrow} F(A) \overset{F(\pi_2)}{\longrightarrow} F(A_2) \to 0$$

is a split exact sequence in $\mathscr{C}_\Delta$.

Proof. Choose $\sigma_2: A_2 \to A, \pi_1: A \to A_1$ so that

$$[\sigma_1, \sigma_2; A; \pi_1, \pi_2] = A_1 * A_2.$$

Then, by Theorem 1, $[F(\sigma_1), F(\sigma_2); F(A); F(\pi_1), F(\pi_2)] = F(A_1) * F(A_2)$ and therefore
$$0 \to F(A_1) \overset{F(\sigma_1)}{\longrightarrow} F(A) \overset{F(\pi_1)}{\longrightarrow} F(A_2) \to 0$$

is a split exact sequence by virtue of Theorem 4.

Exercise 3. *In the diagram*

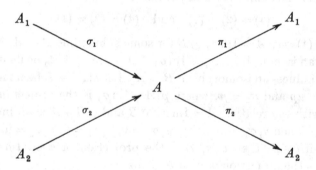

suppose that $\pi_1 \sigma_1 = $ *identity and that* $\pi_2 \sigma_2 = $ *identity. Suppose also that*

$$A_1 \overset{\sigma_1}{\to} A \overset{\pi_2}{\to} A_2 \quad and \quad A_2 \overset{\sigma_2}{\to} A \overset{\pi_1}{\to} A_1$$

are exact. Show that $[\sigma_1, \sigma_2; A; \pi_1, \pi_2] = A_1 * A_2$.

Exercise 4. *Let* $F: \mathscr{C}_\Lambda \to \mathscr{C}_\Delta$ *be a covariant functor and suppose that whenever* $0 \to A' \to A \to A'' \to 0$ *is a split exact sequence in* $\mathscr{C}_\Lambda$, *then* $0 \to F(A') \to F(A) \to F(A'') \to 0$ *is a split exact sequence in* $\mathscr{C}_\Delta$. *Deduce that* F *is additive.*

Let $F: \mathscr{C}_\Lambda \to \mathscr{C}_\Delta$ be a covariant functor. Assume that whenever $0 \to A' \to A \to A'' \to 0$ is exact in $\mathscr{C}_\Lambda$, then

$$0 \to F(A') \to F(A) \to F(A'')$$

resp. $F(A') \to F(A) \to F(A'') \to 0$

is exact in $\mathscr{C}_\Delta$. In these circumstances we say that F is *left exact* resp. *right exact*. Should it be the case that the exactness of

$$0 \to A' \to A \to A'' \to 0$$

only implies that of $\quad F(A') \to F(A) \to F(A''),$

then F is said to be *half exact*. If F is both left and right exact, i.e. if $0 \to A' \to A \to A'' \to 0$ is exact always implies that

$$0 \to F(A') \to F(A) \to F(A'') \to 0$$

is exact, then F is said to be an *exact* functor.

Let $F:\mathscr{C}_\Lambda \to \mathscr{C}_\Delta$ be a covariant functor. If F is left exact then it preserves monomorphisms, whereas if it is right exact it preserves epimorphisms.

Lemma 2. *Suppose that the covariant functor F is left exact and that $0 \to A_1 \to A \to A_2$ is exact in $\mathscr{C}_\Lambda$. Then $0 \to F(A_1) \to F(A) \to F(A_2)$ is exact in $\mathscr{C}_\Delta$.*

The proofs of this and the next two lemmas are straightforward and will be omitted. In both Lemmas 3 and 4, F is understood to be a *covariant* functor from $\mathscr{C}_\Lambda$ to $\mathscr{C}_\Delta$.

Lemma 3. *Suppose that F is right exact and $A_1 \to A \to A_2 \to 0$ is exact in $\mathscr{C}_\Lambda$. Then $F(A_1) \to F(A) \to F(A_2) \to 0$ is exact in $\mathscr{C}_\Delta$.*

Lemma 4. *Suppose that F is exact and $A_1 \to A \to A_2$ is an exact sequence in $\mathscr{C}_\Lambda$. Then $F(A_1) \to F(A) \to F(A_2)$ is exact in $\mathscr{C}_\Delta$.*

Theorem 6. *If the covariant functor F is half exact, then it is additive.*

Proof. Let $[\sigma_1, \sigma_2; A; \pi_1, \pi_2] = A_1 * A_2$ in $\mathscr{C}_\Lambda$. By Theorem 2, it is enough to show that $[F(\sigma_1), F(\sigma_2); F(A); F(\pi_1), F(\pi_2)]$ equals

$$F(A_1) * F(A_2).$$

Now by Theorem 3 and the half exactness of F,

$$F(A_1) \xrightarrow{F(\sigma_1)} F(A) \xrightarrow{F(\pi_2)} F(A_2)$$

and

$$F(A_2) \xrightarrow{F(\sigma_2)} F(A) \xrightarrow{F(\pi_1)} F(A_1)$$

are exact. Consider the diagram

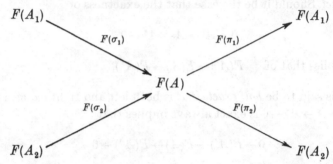

Exercise 3 shows at once that

$$[F(\sigma_1), F(\sigma_2); F(A); F(\pi_1), F(\pi_2)] = F(A_1) * F(A_2).$$

This completes the proof.

Suppose that the covariant functor F is exact and that A_1 is a submodule of the Λ-module A. Then the inclusion mapping gives rise to an exact sequence $0 \to A_1 \to A$ and therefore $0 \to F(A_1) \to F(A)$ is exact. Thus $F(A_1)$ may be regarded as a Δ-submodule of $F(A)$. This observation is relevant to the next two exercises.

Exercise 5. *Let $F: \mathscr{C}_\Lambda \to \mathscr{C}_\Delta$ be a covariant exact functor. Show that F preserves images and kernels.*

Exercise 6. *Suppose that the functor $F: \mathscr{C}_\Lambda \to \mathscr{C}_\Delta$ is exact and covariant, that A is a Λ-module, and $A_1, A_2, ..., A_n$ are Λ-submodules of A. Show that, as submodules of $F(A)$,*

$$F(A_1 + A_2 + ... + A_n) = F(A_1) + F(A_2) + ... + F(A_n)$$

and $F(A_1 \cap A_2 \cap ... \cap A_n) = F(A_1) \cap F(A_2) \cap ... \cap F(A_n).$

1.4 Contravariant functors

We now introduce a second type of functor. Assume that with each module A in $\mathscr{C}_\Lambda$ there is associated a module $G(A)$ in $\mathscr{C}_\Delta$ and that with each Λ-homomorphism $f: A \to A'$ there is associated a Δ-homomorphism $G(f): G(A') \to G(A)$. Assume further that

(1) $G(i_A) = i_{G(A)}$ *for all A in $\mathscr{C}_\Lambda$;*

(2) $G(gf) = G(f) G(g)$ *whenever $f: A \to A'$ and $g: A' \to A''$ in $\mathscr{C}_\Lambda$.*

We then say that we have a *contravariant functor* G from Λ-modules to Δ-modules.

Let $G:\mathscr{C}_\Lambda \to \mathscr{C}_\Delta$ be a contravariant functor. If

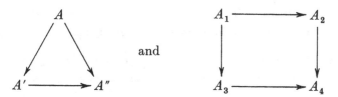

are commutative diagrams in $\mathscr{C}_\Lambda$, then applying G leaves them commutative but the arrows are reversed. Also if $f:A \to A'$ is an isomorphism and $g:A' \to A$ is its inverse, then $G(f):G(A') \to G(A)$ is an isomorphism and $G(g):G(A) \to G(A')$ is its inverse. It is clear how *additive* contravariant functors are defined. If G is additive, then it converts null homomorphisms and null objects into null homomorphisms and null objects.

Theorem 7. *Let $G:\mathscr{C}_\Lambda \to \mathscr{C}_\Delta$ be an additive contravariant functor and let $[\sigma_1, ..., \sigma_n; A; \pi_1, ..., \pi_n] = A_1 * ... * A_n$ in $\mathscr{C}_\Lambda$. Then*

$$[G(\pi_1), ..., G(\pi_n); G(A); G(\sigma_1), ..., G(\sigma_n)] = G(A_1) * G(A_2) * ... * G(A_n)$$
in $\mathscr{C}_\Delta$.

Proof. Apply G to the relations $\pi_j \sigma_i = \delta_{ji}$ and $\Sigma \sigma_i \pi_i =$ identity.

Theorem 8. *Let $G:\mathscr{C}_\Lambda \to \mathscr{C}_\Delta$ be a contravariant functor and suppose that whenever $[\sigma_1, \sigma_2; A; \pi_1, \pi_2] = A_1 * A_2$ in $\mathscr{C}_\Lambda$, then*

$$[G(\pi_1), G(\pi_2); G(A); G(\sigma_1), G(\sigma_2)] = G(A_1) * G(A_2)$$

in $\mathscr{C}_\Delta$. In these circumstances G is additive.

Proof. Suppose that $f_1, f_2:A \to B$ and let $[\sigma_1, \sigma_2; X; \pi_1, \pi_2] = B * B$. Then $\pi_1 + \pi_2:X \to B$. Let $g:A \to X$ be such that $\pi_\mu g = f_\mu$, and therefore $(\pi_1 + \pi_2)g = \pi_1 g + \pi_2 g = f_1 + f_2$. By hypothesis,

$$[G(\pi_1), G(\pi_2); G(X); G(\sigma_1), G(\sigma_2)] = G(B) * G(B).$$

Furthermore $\qquad G(\pi_1 + \pi_2):G(B) \to G(X)$

and $\qquad G(\sigma_\mu)\, G(\pi_1 + \pi_2) = G(\overline{\pi_1 + \pi_2}\, \sigma_\mu) = i_{G(B)}.$

In addition $\qquad G(\pi_1) + G(\pi_2):G(B) \to G(X),$

and $\qquad G(\sigma_\mu)\, (G(\pi_1) + G(\pi_2)) = i_{G(B)}$

as well. It follows (see Exercise 1) that $G(\pi_1+\pi_2) = G(\pi_1)+G(\pi_2)$. Finally

$$G(f_1+f_2) = G(\overline{\pi_1+\pi_2}\,g) = G(g)\,G(\pi_1+\pi_2) = G(g)\,(G(\pi_1)+G(\pi_2))$$
$$= G(g)\,G(\pi_1)+G(g)\,G(\pi_2) = G(\pi_1 g)+G(\pi_2 g) = G(f_1)+G(f_2).$$

Theorem 9. *Let $0 \to A_1 \xrightarrow{\sigma_1} A \xrightarrow{\pi_2} A_2 \to 0$ be a split exact sequence in $\mathscr{C}_\Lambda$ and let $G: \mathscr{C}_\Lambda \to \mathscr{C}_\Delta$ be an additive contravariant functor. Then*

$$0 \to G(A_2) \xrightarrow{G(\pi_2)} G(A) \xrightarrow{G(\sigma_1)} G(A_1) \to 0$$

is a split exact sequence in $\mathscr{C}_\Delta$.

Proof. This is the analogue of Theorem 5 and we simply modify the proof of that result.

Let $G: \mathscr{C}_\Lambda \to \mathscr{C}_\Delta$ be a contravariant functor. It is said to be *left exact* if whenever $0 \to A' \to A \to A'' \to 0$ is exact in $\mathscr{C}_\Lambda$, then

$$0 \to G(A'') \to G(A) \to G(A') \tag{1.4.1}$$

is exact in $\mathscr{C}_\Delta$. We say that G is *right exact* if, in place of (1.4.1), we have an exact sequence

$$G(A'') \to G(A) \to G(A') \to 0. \tag{1.4.2}$$

Furthermore G is called *half exact* if the exactness of

$$0 \to A' \to A \to A'' \to 0$$

only implies that of $G(A'') \to G(A) \to G(A').$ \hfill (1.4.3)

Finally G is said to be *exact* if it is both left exact and right exact, i.e. if the exactness of the sequence $0 \to A' \to A \to A'' \to 0$ implies that

$$0 \to G(A'') \to G(A) \to G(A') \to 0 \tag{1.4.4}$$

is exact.

Lemma 5. *Suppose that $A' \to A \to A'' \to 0$ is exact in $\mathscr{C}_\Lambda$ and that $G: \mathscr{C}_\Lambda \to \mathscr{C}_\Delta$ is contravariant and left exact. Then*

$$0 \to G(A'') \to G(A) \to G(A')$$

is exact in $\mathscr{C}_\Delta$.

This is trivial as are Lemmas 6 and 7 provided that one bears in mind that a left exact contravariant functor converts an epimorphism into a monomorphism whereas a right exact contravariant functor changes a monomorphism into an epimorphism.

Lemma 6. *Suppose that $G: \mathscr{C}_\Lambda \to \mathscr{C}_\Delta$ is contravariant and right exact. If now $0 \to A' \to A \to A''$ is exact in $\mathscr{C}_\Lambda$, then $G(A'') \to G(A) \to G(A') \to 0$ is exact in $\mathscr{C}_\Lambda$.*

Lemma 7. *Suppose that the functor* $G:\mathscr{C}_\Lambda \to \mathscr{C}_\Delta$ *is contravariant and exact. If now* $A' \to A \to A''$ *is exact in* $\mathscr{C}_\Lambda$, *then* $G(A'') \to G(A) \to G(A')$ *is exact in* $\mathscr{C}_\Delta$.

Theorem 10. *Let* G *be a half exact contravariant functor. Then it is additive.*

The proof is similar to that of Theorem 6 save that we use Theorem 8 in place of Theorem 2.

1.5 Additional structure

Let $F:\mathscr{C}_\Lambda \to \mathscr{C}_\Delta$ be an *additive* covariant functor. Suppose that A is a (Λ, Γ)-bimodule. For definiteness we shall assume that A belongs to $\mathscr{C}_\Gamma^L$. If $\gamma \in \Gamma$, then multiplication by γ induces a Λ-homomorphism $A \overset{\gamma}{\to} A$ and this will give rise to a Δ-homomorphism $F(A) \to F(A)$. For $x \in F(A)$, let us define γx to be the image of x under $F(A) \to F(A)$. If now $\gamma_1, \gamma_2 \in \Gamma$, then

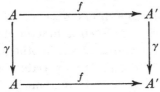

$$(1.5.1)$$

is a commutative diagram in $\mathscr{C}_\Lambda$ and it follows, on applying F, that $\gamma_2(\gamma_1 x) = (\gamma_2 \gamma_1) x$. In fact, because F is *additive*, $F(A)$ is now a (Δ, Γ)-bimodule with $F(A)$ in $\mathscr{C}_\Gamma^L$. Also if A, A' are (Λ, Γ)-bimodules and $f:A \to A'$ is a bihomomorphism, then because the diagram

$$
\begin{array}{ccc}
A & \overset{f}{\longrightarrow} & A' \\
\gamma \downarrow & & \downarrow \gamma \\
A & \overset{f}{\longrightarrow} & A'
\end{array}
$$

is commutative, so too is

$$
\begin{array}{ccc}
F(A) & \overset{F(f)}{\longrightarrow} & F(A') \\
\gamma \downarrow & & \downarrow \gamma \\
F(A) & \overset{F(f)}{\longrightarrow} & F(A')
\end{array}
$$

Thus $F(f)$ is a (Δ, Γ)-bihomomorphism.

We have supposed that A belongs to $\mathscr{C}_\Gamma^L$. If instead we had assumed that A was a (Λ, Γ)-bimodule with A in $\mathscr{C}_\Gamma^R$, then the conclusions would have been the same except that $F(A)$ would have had the structure of a (Δ, Γ)-bimodule with $F(A)$ in $\mathscr{C}_\Gamma^R$.

Next assume that $G: \mathscr{C}_\Lambda \to \mathscr{C}_\Delta$ is a *contravariant additive* functor. To begin with suppose that A is a (Λ, Γ)-bimodule with A in $\mathscr{C}_\Gamma^L$. If now $x \in G(A)$ and $\gamma \in \Gamma$, define $x\gamma$ to be the image of x under the mapping $G(A) \to G(A)$ induced by $A \overset{\gamma}{\to} A$. The commutative diagram (1.5.1) now yields

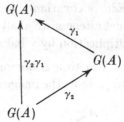

which is also commutative, so $(x\gamma_2)\gamma_1 = x(\gamma_2\gamma_1)$. In fact $G(A)$ is a (Δ, Γ)-bimodule with $G(A)$ in $\mathscr{C}_\Gamma^R$, and each bihomomorphism $A \to A'$ induces a bihomomorphism $G(A') \to G(A)$. (If we suppose that A is a (Λ, Γ)-bimodule with A in $\mathscr{C}_\Gamma^R$, then $G(A)$ will be a (Δ, Γ)-bimodule with $G(A)$ in $\mathscr{C}_\Gamma^L$.) Note that *in the contravariant case, the side on which the auxiliary ring Γ operates is changed on applying the functor.*

1.6 Bifunctors

Besides considering functors in one variable, we may also consider functors in many variables. Such a functor may be covariant in some of its variables and contravariant in others. By way of illustration we shall discuss a functor of two variables, i.e. a *bifunctor*, contravariant in the first variable and covariant in the second.

Suppose that with each module A in $\mathscr{C}_\Lambda$ and B in $\mathscr{C}_\Gamma$ there is associated a module $T(A, B)$ in $\mathscr{C}_\Delta$, and suppose also that given $f: A' \to A$ in $\mathscr{C}_\Lambda$ and $g: B \to B'$ in $\mathscr{C}_\Gamma$, there corresponds a Δ-homomorphism $T(f, g): T(A, B) \to T(A', B')$. If now

(1) $T(i_A, i_B) = i_{T(A, B)}$,

(2) $T(ff_1, g_1g) = T(f_1, g_1) T(f, g)$ *whenever* $A'' \overset{f_1}{\to} A' \overset{f}{\to} A$ *in* $\mathscr{C}_\Lambda$ *and* $B \overset{g}{\to} B' \overset{g_1}{\to} B''$ *in* $\mathscr{C}_\Gamma$,

then we say that T is a *bifunctor* from $\mathscr{C}_\Lambda \times \mathscr{C}_\Gamma$ to $\mathscr{C}_\Delta$ which is *contravariant in the first variable and covariant in the second*.

Assume that T is such a functor. If we keep A *fixed*, then we obtain a covariant functor $T(A,-):\mathscr{C}_\Gamma \to \mathscr{C}_\Delta$. Here it is understood that if $g:B \to B'$ in $\mathscr{C}_\Gamma$, then by $T(A,g)$ we mean $T(i_A,g)$. Likewise if B is kept fixed, then we obtain a contravariant functor $T(-,B):\mathscr{C}_\Lambda \to \mathscr{C}_\Delta$. If both $T(A,-)$ and $T(-,B)$ are additive (for all choices of A and B), then T is said to be *additive*. Thus if T is additive, then

$$T(A,g_1+g_2) = T(A,g_1) + T(A,g_2)$$

when $g_1, g_2:B \to B'$ in $\mathscr{C}_\Gamma$, and

$$T(f_1+f_2, B) = T(f_1, B) + T(f_2, B)$$

provided that $f_1, f_2:A' \to A$ in $\mathscr{C}_\Lambda$. Again, if both $T(A,-)$ and $T(-,B)$ are left exact (for all choices of A and B), then T is called *left exact*. Accordingly if T is left exact and the sequences

$$A_1 \to A \to A_2 \to 0$$

and $$0 \to B_1 \to B \to B_2$$

are exact in $\mathscr{C}_\Lambda$ and $\mathscr{C}_\Gamma$ respectively, then both

$$0 \to T(A_2, B) \to T(A, B) \to T(A_1, B)$$

and $$0 \to T(A, B_1) \to T(A, B) \to T(A, B_2)$$

will be exact in $\mathscr{C}_\Delta$. The notions of right exact, exact, and half exact functors are generalized in the same way.

We continue to suppose that $T(A,B)$ is contravariant in A and covariant in B but now add the assumption that T is *additive*. Let Ω be a ring with identity and suppose that A is a (Λ, Ω)-bimodule. Then $T(A,B)$ is a (Δ, Ω)-bimodule. However, because $T(-,B)$ is *contravariant*, Ω will operate on *different* sides in the two cases so that if, for example, A belongs to $\mathscr{C}_\Omega^L$, then $T(A,B)$ will belong to $\mathscr{C}_\Omega^R$. We know that a bihomomorphism $f:A' \to A$ of (Λ, Ω)-bimodules induces a bihomomorphism $T(A,B) \to T(A', B)$. Still assuming that A is a bimodule, suppose that $g:B \to B'$ in $\mathscr{C}_\Gamma$ and that $\omega \in \Omega$. Then, with a self-explanatory notation,

$$A \overset{\omega}{\to} A \overset{i_A}{\to} A = A \overset{i_A}{\to} A \overset{\omega}{\to} A$$

and $$B \overset{g}{\to} B' \overset{i_{B'}}{\to} B' = B \overset{i_B}{\to} B \overset{g}{\to} B'.$$

Consequently $\quad T(\omega, i_{B'})\, T(i_A, g) = T(i_A, g)\, T(\omega, i_B)$
which shows that the diagram

$$
\begin{array}{ccc}
T(A, B) & \xrightarrow{\;\;T(A,\,g)\;\;} & T(A, B') \\
\downarrow{\scriptstyle\omega} & & \downarrow{\scriptstyle\omega} \\
T(A, B) & \xrightarrow{\;\;T(A,\,g)\;\;} & T(A, B')
\end{array}
$$

is commutative. It follows that $T(A, g): T(A, B) \to T(A, B')$ is a *bi-homomorphism* as well.

Now assume that B is a (Γ, Ω)-bimodule. Then $T(A, B)$ is a (Δ, Ω)-bimodule and Ω acts on the same side of $T(A, B)$ as it does in the case of B. Each bihomomorphism $B \to B'$ of (Γ, Ω)-bimodules induces a bihomomorphism $T(A, B) \to T(A, B')$. Also each Λ-homomorphism $f: A' \to A$ induces a bihomomorphism

$$ T(f, B): T(A, B) \to T(A', B). $$

1.7 Equivalent functors

The notions introduced in this section apply to functors in arbitrarily many variables, but we shall explain them by considering primarily functors of two variables, where the functors concerned are contravariant in the first variable and covariant in the second.

Let $T(A, B)$ and $U(A, B)$ be functors from $\mathscr{C}_\Lambda \times \mathscr{C}_\Gamma$ to $\mathscr{C}_\Delta$ which are contravariant in A and covariant in B. Suppose that for each module A in $\mathscr{C}_\Lambda$ and each module B in $\mathscr{C}_\Gamma$ there is defined a Δ-homomorphism

$$ \eta_{AB}: T(A, B) \to U(A, B) \tag{1.7.1} $$

and that these are such that, whenever $A' \to A$ in $\mathscr{C}_\Lambda$ and $B \to B'$ in $\mathscr{C}_\Gamma$, the diagram

$$
\begin{array}{ccc}
T(A, B) & \xrightarrow{\;\;\eta_{AB}\;\;} & U(A, B) \\
\downarrow & & \downarrow \\
T(A', B') & \xrightarrow{\;\;\eta_{A'B'}\;\;} & U(A', B')
\end{array}
$$

is commutative, i.e. we assume that the homomorphism (1.7.1) is *natural* for homomorphisms $A' \to A$ and $B \to B'$. We then say that we have a *natural transformation* $\eta: T \to U$.

For example, each Γ-homomorphism $B \to B'$ induces a natural transformation $T(-, B) \to T(-, B')$ between $T(-, B)$ and $T(-, B')$ considered as functors of a single variable which varies in $\mathscr{C}_\Lambda$. Likewise each Λ-homomorphism $A' \to A$ induces a natural transformation $T(A, -) \to T(A', -)$ between $T(A, -)$ and $T(A', -)$ considered as functors of a single variable which this time varies in $\mathscr{C}_\Gamma$.

Natural transformations of functors may be composed in an obvious manner.

Finally suppose that the natural transformation $\eta: T \to U$ is such that $\eta_{AB}: T(A, B) \to U(A, B)$ is an *isomorphism* in $\mathscr{C}_\Lambda$ for every A in $\mathscr{C}_\Lambda$ and B in $\mathscr{C}_\Gamma$. In this situation η is called a *natural equivalence* and we say that T and U are *naturally equivalent*. The relation of natural equivalence between functors is reflexive, symmetric and transitive.

Example. Let I be the *identity functor* on $\mathscr{C}_\Lambda$ so that for every Λ-module A and Λ-homomorphism f, $I(A) = A$ and $I(f) = f$. A natural transformation $\eta: I \to I$ is then characterized by the fact that whenever $f: A \to A'$ in $\mathscr{C}_\Lambda$, the diagram

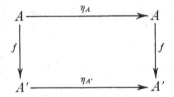

is commutative. If η and η' are two such natural transformations, then they may be 'added' and 'multiplied' in such a way that

$$(\eta + \eta')_A = \eta_A + \eta'_A \ \text{ and } \ (\eta \eta')_A = \eta_A \eta'_A.$$

Note that if, in the above diagram, we take f to be $\eta'_A: A \to A$, then we find that $\eta \eta' = \eta' \eta$, i.e. our multiplication is commutative.

Let γ belong to the centre of Λ. We obtain a natural transformation $\eta: I \to I$ by taking η_A to consist of multiplication by γ. The next exercise shows, among other things, that all natural transformations of I into itself arise in this way.

Exercise 7. *Establish a bijection between the elements of the centre of Λ and the natural transformations of the identity functor (on $\mathscr{C}_\Lambda$) into itself. Deduce that the natural transformations of the identity functor into itself form a ring which is isomorphic to the centre of Λ.*

A point has now been reached where enough of the language of functors has been developed for our immediate purposes. To save time we have worked exclusively with modules, but of course it is possible to prove similar results in a much broader context. The reader who wishes to see how this extra generality can be achieved may like to consult (7) and (17) in the list of references provided at the end.

Solutions to the Exercises on Chapter 1

Exercise 1. *Let* $[\sigma_1, ..., \sigma_n; A; \pi_1, ..., \pi_n] = A_1 * A_2 * ... * A_n$ *in* $\mathscr{C}_\Lambda$. *Show that if* Λ*-homomorphisms* $f_i: A_i \to B$ $(1 \leqslant i \leqslant n)$ *are given, then there exists a unique homomorphism* $f: A \to B$ *such that* $f\sigma_i = f_i$ *for* $1 \leqslant i \leqslant n$. *Show also that if* $g_i: B \to A_i$ $(1 \leqslant i \leqslant n)$ *are prescribed* Λ*-homomorphisms, then there exists a unique homomorphism* $g: B \to A$ *such that* $\pi_i g = g_i$ *for* $1 \leqslant i \leqslant n$.

Solution. Define a homomorphism $f: A \to B$ by $f = \sum_{j=1}^{n} f_j \pi_j$. Then

$$f\sigma_i = \left(\sum_{j=1}^{n} f_j \pi_j \right) \sigma_i = \sum_{j=1}^{n} f_j \pi_j \sigma_i = \sum_{j=1}^{n} f_j \delta_{ji} = f_i \quad (1 \leqslant i \leqslant n).$$

Let $f': A \to B$ be another homomorphism such that $f'\sigma_i = f_i (1 \leqslant i \leqslant n)$.

Then $\quad f' = f' i_A = f' \left(\sum_{i=1}^{n} \sigma_i \pi_i \right) = \sum_{i=1}^{n} f'\sigma_i \pi_i = \sum_{i=1}^{n} f_i \pi_i = f.$

This proves uniqueness.

Now define a homomorphism $g: B \to A$ by $g = \sum_{j=1}^{n} \sigma_j g_j$. Then

$$\pi_i g = \pi_i \left(\sum_{j=1}^{n} \sigma_j g_j \right) = \sum_{j=1}^{n} \pi_i \sigma_j g_j = \sum_{j=1}^{n} \delta_{ij} g_j = g_i \quad (1 \leqslant i \leqslant n).$$

Let $g': B \to A$ be another homomorphism such that $\pi_i g' = g_i (1 \leqslant i \leqslant n)$. Then
$$g' = i_A g' = \left(\sum_{i=1}^{n} \sigma_i \pi_i \right) g' = \sum_{i=1}^{n} \sigma_i \pi_i g' = \sum_{i=1}^{n} \sigma_i g_i = g.$$

This completes the solution.

Exercise 2. *Let* $[\sigma_1, \sigma_2, ..., \sigma_n; A; \pi_1, \pi_2, ..., \pi_n] = A_1 * A_2 * ... * A_n$ *in* $\mathscr{C}_\Lambda$. *Show that the homomorphism* $A_1 \oplus A_2 \oplus ... \oplus A_n \to A$ *induced by the* σ_i *and the homomorphism* $A \to A_1 \times A_2 \times ... \times A_n$ *induced by the* π_i *are isomorphisms.*

Solution. Let $\phi: A_1 \oplus A_2 \oplus ... \oplus A_n \to A$ be the homomorphism induced by the σ_i and $\psi: A \to A_1 \times A_2 \times ... \times A_n$ the homomorphism induced by the π_i. Then

$$A_1 \times A_2 \times ... \times A_n = A_1 \oplus A_2 \oplus ... \oplus A_n, \phi((a_1, a_2, ..., a_n)) = \sum_{j=1}^{n} \sigma_j(a_j)$$
and $\qquad\qquad \psi(a) = (\pi_1(a), \pi_2(a), ..., \pi_n(a)).$

We have $\phi\psi : A \to A$ and, in fact, for all $a \in A$

$$\phi\psi(a) = \phi((\pi_1(a), \pi_2(a), ..., \pi_n(a)))$$

$$= \sum_{j=1}^{n} \sigma_j \pi_j(a) = \left(\sum_{j=1}^{n} \sigma_j \pi_j \right)(a) = i_A(a).$$

Consequently $\phi\psi = i_A$. Also

$$\phi\psi : A_1 \oplus A_2 \oplus ... \oplus A_n \to A_1 \oplus A_2 \oplus ... \oplus A_n$$

and, for all $(a_1, a_2, ..., a_n)$ in $A_1 \oplus A_2 \oplus ... \oplus A_n$, we have

$$\psi\phi((a_1, a_2, ..., a_n)) = \psi\left(\sum_{j=1}^{n} \sigma_j(a_j) \right)$$

$$= \left(\pi_1 \left(\sum_{j=1}^{n} \sigma_j(a_j) \right), \pi_2 \left(\sum_{j=1}^{n} \sigma_j(a_j) \right), ..., \pi_n \left(\sum_{j=1}^{n} \sigma_j(a_j) \right) \right)$$

$$= \left(\sum_{j=1}^{n} \pi_1 \sigma_j(a_j), \sum_{j=1}^{n} \pi_2 \sigma_j(a_j), ..., \sum_{j=1}^{n} \pi_n \sigma_j(a_j) \right)$$

$$= \left(\sum_{j=1}^{n} \delta_{1j}(a_j), \sum_{j=1}^{n} \delta_{2j}(a_j), ..., \sum_{j=1}^{n} \delta_{nj}(a_j) \right)$$

$$= (a_1, a_2, ..., a_n).$$

Thus $\psi\phi$ is also an identity map. Accordingly ϕ and ψ are isomorphisms and furthermore $\psi = \phi^{-1}$.

Exercise 3. *In the diagram*

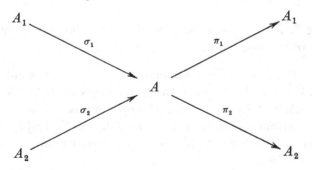

suppose that $\pi_1\sigma_1 = $ identity and that $\pi_2\sigma_2 = $ identity. Suppose also that

$$A_1 \overset{\sigma_1}{\to} A \overset{\pi_2}{\to} A_2 \quad and \quad A_2 \overset{\sigma_2}{\to} A \overset{\pi_1}{\to} A_1$$

are exact. Show that $[\sigma_1, \sigma_2; A; \pi_1, \pi_2] = A_1 * A_2$.

Solution. Since $A_1 \overset{\sigma_1}{\to} A \overset{\pi_2}{\to} A_2$ and $A_2 \overset{\sigma_2}{\to} A \overset{\pi_1}{\to} A_1$ are exact, $\pi_2 \sigma_1 = 0$ and $\pi_1 \sigma_2 = 0$. We therefore obtain $\pi_i \sigma_j = \delta_{ij}$ for $i, j = 1, 2$. To prove that $[\sigma_1, \sigma_2; A; \pi_1, \pi_2] = A_1 * A_2$ it remains to be shown that

$$i_A = \sigma_1 \pi_1 + \sigma_2 \pi_2.$$

Now by Lemma 1, $A = \operatorname{Im} \sigma_1 \oplus \operatorname{Ker} \pi_1$. However $\operatorname{Ker} \pi_1 = \operatorname{Im} \sigma_2$ and therefore $A = \operatorname{Im} \sigma_1 \oplus \operatorname{Im} \sigma_2$. Let $a \in A$. Then $a = \sigma_1(a_1) + \sigma_2(a_2)$, where $a_i \in A_i$. It follows that $\pi_i(a) = a_i$. Consequently

$$(\sigma_1 \pi_1 + \sigma_2 \pi_2)(a) = \sigma_1(a_1) + \sigma_2(a_2) = a.$$

The desired result follows.

Exercise 4. *Let $F: \mathscr{C}_\Lambda \to \mathscr{C}_\Delta$ be a covariant functor and suppose that whenever $0 \to A' \to A \to A'' \to 0$ is a split exact sequence in $\mathscr{C}_\Lambda$, then $0 \to F(A') \to F(A) \to (A'') \to 0$ is a split exact sequence in $\mathscr{C}_\Delta$. Deduce that F is additive.*

Solution. Let $[\sigma_1, \sigma_2; A; \pi_1, \pi_2] = A_1 * A_2$ be an arbitrary biproduct. Then $0 \to A_1 \overset{\sigma_1}{\to} A \overset{\pi_2}{\to} A_2 \to 0$ is a split exact sequence, and therefore $0 \to F(A_1) \overset{F(\sigma_1)}{\longrightarrow} F(A) \overset{F(\pi_2)}{\longrightarrow} F(A_2) \to 0$ is split exact as well. In particular $F(A_1) \overset{F(\sigma_1)}{\longrightarrow} F(A) \overset{F(\pi_2)}{\longrightarrow} F(A_2)$ is exact. Similarly $F(A_2) \overset{F(\sigma_2)}{\longrightarrow} F(A) \overset{F(\pi_1)}{\longrightarrow} F(A_1)$ is exact. Next $F(\pi_i) F(\sigma_i) = F(\pi_i \sigma_i) = $ identity. Hence, by Exercise 3, $[F(\sigma_1), F(\sigma_2); F(A); F(\pi_1), F(\pi_2)] = F(A_1) * F(A_2)$, and therefore, by Theorem 2, F is additive.

Exercise 5. *Let $F: \mathscr{C}_\Lambda \to \mathscr{C}_\Delta$ be a covariant exact functor. Show that F preserves images and kernels.*

Solution. Let $f: A \to B$ be a homomorphism. Then f is composed of the epimorphism $A \to \operatorname{Im} f$ and the monomorphism $\operatorname{Im} f \to B$. Hence $F(f)$ is obtained by combining the epimorphism $F(A) \to F(\operatorname{Im} f)$ with the monomorphism $F(\operatorname{Im} f) \to F(B)$. Thus $\operatorname{Im} F(f)$ is just $F(\operatorname{Im} f)$ regarded as a submodule of $F(B)$. We may therefore write symbolically $\operatorname{Im} F(f) = F(\operatorname{Im} f)$.

Next, since $\operatorname{Ker} f \to A \to B$ is exact, so too is $F(\operatorname{Ker} f) \to F(A) \to F(B)$ and, moreover, $F(\operatorname{Ker} f) \to F(A)$ is monic. Hence

$$\operatorname{Ker} F(f) = F(\operatorname{Ker} f)$$

provided that the latter is regarded as a submodule of $F(A)$.

Exercise 6. *Suppose that the functor $F: \mathscr{C}_\Lambda \to \mathscr{C}_\Delta$ is exact and covariant, that A is a Λ-module, and $A_1, A_2, \ldots, A_n$ are Λ-submodules of A. Show that, as submodules of $F(A)$,*

$$F(A_1 + A_2 + \ldots + A_n) = F(A_1) + F(A_2) + \ldots + F(A_n)$$

and $\quad F(A_1 \cap A_2 \cap \ldots \cap A_n) = F(A_1) \cap F(A_2) \cap \ldots \cap F(A_n).$

Solution. We may suppose that $n = 2$. Let B_1, B_2 be submodules of A with $B_1 \subseteq B_2$. On applying F to the commutative diagram

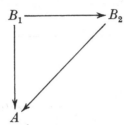

formed by the various inclusion mappings, we see that $F(B_1) \subseteq F(B_2)$. Now $A_i \subseteq A_1 + A_2$ and $A_1 \cap A_2 \subseteq A_i$. Thus $F(A_i) \subseteq F(A_1 + A_2)$ and $F(A_1 \cap A_2) \subseteq F(A_i)$. Accordingly $F(A_1) + F(A_2) \subseteq F(A_1 + A_2)$ and $F(A_1 \cap A_2) \subseteq F(A_1) \cap F(A_2)$.

Let $X = A_1 \oplus A_2$ and define $f_i: X \to A$ by $f_i(a_1, a_2) = a_i$. Then

$$(f_1 + f_2)(a_1, a_2) = a_1 + a_2.$$

Consequently $\operatorname{Im} f_i = A_i$ and $\operatorname{Im}(f_1 + f_2) = A_1 + A_2$. But F is additive (Theorem 6). Hence, by Exercise 5, we have (as submodules of $F(A)$)

$$\begin{aligned}
F(A_1 + A_2) &= F(\operatorname{Im}(f_1 + f_2)) = \operatorname{Im}(F(f_1 + f_2)) \\
&= \operatorname{Im}(F(f_1) + F(f_2)) \subseteq \operatorname{Im} F(f_1) + \operatorname{Im} F(f_2) \\
&= F(\operatorname{Im} f_1) + F(\operatorname{Im} f_2) = F(A_1) + F(A_2).
\end{aligned}$$

It follows that $F(A_1 + A_2) = F(A_1) + F(A_2)$.

Let $Y = A/A_1 \times A/A_2$. Define $g_i: A \to Y \; (i = 1, 2)$ by

$$g_1(a) = \{\phi_1(a), 0\}, g_2(a) = \{0, \phi_2(a)\},$$

where $\phi_i: A \to A/A_i$ is the natural mapping. Then

$$(g_1 + g_2)(a) = \{\phi_1(a), \phi_2(a)\}.$$

Accordingly $\operatorname{Ker} g_i = A_i$ and $\operatorname{Ker}(g_1 + g_2) = A_1 \cap A_2$. Thus, as submodules of $F(A)$,

$$\begin{aligned}
F(A_1 \cap A_2) &= F(\operatorname{Ker}(g_1 + g_2)) = \operatorname{Ker}(F(g_1 + g_2)) \quad \text{(by Exercise 5)} \\
&= \operatorname{Ker}(F(g_1) + F(g_2)) \supseteq \operatorname{Ker} F(g_1) \cap \operatorname{Ker} F(g_2) \\
&= F(\operatorname{Ker} g_1) \cap F(\operatorname{Ker} g_2) \quad \text{(by Exercise 5)} \\
&= F(A_1) \cap F(A_2).
\end{aligned}$$

Accordingly $F(A_1 \cap A_2) = F(A_1) \cap F(A_2)$.

Exercise 7. *Establish a bijection between the elements of the centre of* Λ *and the natural transformations of the identity functor (on* $\mathscr{C}_\Lambda$*) into itself. Deduce that the natural transformations of the identity functor into itself form a ring which is isomorphic to the centre of* Λ.

Solution. We need only consider left Λ-modules. Let γ be an element of the centre of Λ. For each A in $\mathscr{C}_\Lambda^L$, define a mapping $\eta(\gamma)_A : A \to A$ by $\eta(\gamma)_A(a) = \gamma a$. Since γ is in the centre of Λ, $\eta(\gamma)_A$ is a Λ-homomorphism. Also if $f : A \to B$ is in $\mathscr{C}_\Lambda^L$, then the diagram

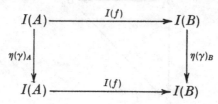

where I is the identity functor, is commutative. It follows that $\eta(\gamma)$ is a natural transformation of I into itself. Hence η maps elements of the centre of Λ to natural transformations of the identity functor into itself.

Let γ_1 and γ_2 be elements of the centre of Λ and assume that $\eta(\gamma_1) = \eta(\gamma_2)$. Then $\eta(\gamma_1)_\Lambda = \eta(\gamma_2)_\Lambda$, where Λ is considered as a left Λ-module, and in particular $\gamma_1 = \eta(\gamma_1)_\Lambda (1) = \eta(\gamma_2)_\Lambda (1) = \gamma_2$. Hence η is an injection.

Next let $\mu : I \to I$ be a natural transformation and put $\gamma = \mu_\Lambda(1)$. If now $\lambda \in \Lambda$, define a homomorphism $f : \Lambda \to \Lambda$ by $f(\lambda') = \lambda'\lambda$. The commutative diagram

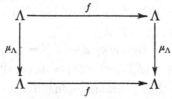

shows that $\gamma\lambda = f(\gamma) = f\mu_\Lambda(1) = \mu_\Lambda(f(1)) = f(1)\mu_\Lambda(1) = \lambda\gamma$. Thus γ is in the centre of Λ. Now assume that A belongs to $\mathscr{C}_\Lambda^L$ and let $a \in A$. Define a homomorphism $g : \Lambda \to A$ by $g(\lambda) = \lambda a$. From the commutative diagram

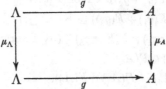

we obtain $\mu_A(a) = \mu_A g(1) = g\mu_\Lambda(1) = g(\gamma) = \gamma a$. Consequently

$$\mu_A = \eta(\gamma)_A$$

and therefore, since A is arbitrary, $\mu = \eta(\gamma)$. Accordingly η is surjective and hence bijective.

Since η is a bijection and the centre of Λ is a subring of Λ, it only remains to be shown that η preserves sums and products. This however is clear.

Supplementary Exercises on Chapter 1

Exercise A. *Let* $\mu: A \to A'$ *be a* Λ-*homomorphism. Show that the following statements are equivalent*:

(1) $\mu(a_1) \neq \mu(a_2)$ *whenever* a_1, a_2 *are distinct elements of* A;

(2) $\mu f_1 = \mu f_2$ *(for* Λ-*homomorphisms* f_1 *and* f_2) *always implies* $f_1 = f_2$.

Solution. Assume (1). Let $f_1, f_2 : B \to A$ be Λ-homomorphisms such that $\mu f_1 = \mu f_2$. Then, for every $b \in B$, $\mu(f_1(b)) = \mu(f_2(b))$ and so, by (1), $f_1(b) = f_2(b)$. Thus $f_1 = f_2$.

Assume (2). We may suppose that we are dealing with left Λ-modules. Let $a_1, a_2 \in A$ be such that $a_1 \neq a_2$. For $i = 1, 2$ define Λ-homomorphisms $f_i : \Lambda \to A$ by $f_i(\lambda) = \lambda a_i$. Then $f_1 \neq f_2$ and therefore, by (b), $\mu f_1 \neq \mu f_2$. Hence there exists $\lambda \in \Lambda$ such that $\mu f_1(\lambda) \neq \mu f_2(\lambda)$, i.e. $\lambda\mu(a_1) \neq \lambda\mu(a_2)$. It follows that $\mu(a_1) \neq \mu(a_2)$.

Exercise B. *Let* $v: A \to A'$ *be a* Λ-*homomorphism. Show that the following statements are equivalent*:

(1) v *is surjective*;

(2) $g_1 v = g_2 v$ *(for* Λ-*homomorphisms* g_1, g_2) *always implies* $g_1 = g_2$.

Solution. Assume (1). Let $g_1, g_2 : A' \to B$ be Λ-homomorphisms such that $g_1 v = g_2 v$. For every $a' \in A'$ there exists $a \in A$ such that $v(a) = a'$. Consequently $g_1(a') = g_1 v(a) = g_2 v(a) = g_2(a')$. Hence $g_1 = g_2$.

Assume (2) and suppose that v is not surjective. Then there exists $a' \in A'$ such that $a' \notin \operatorname{Im} v$. Define $g_i : A' \to A'/\operatorname{Im} v$ $(i = 1, 2)$ by $g_1 = 0$ and g_2 is the natural epimorphism. Then $g_1(a') = 0, g_2(a') \neq 0$ which shows that $g_1 \neq g_2$. But $g_1 v = 0 = g_2 v$, and now (2) gives a contradiction.

Exercise C. *Let* $\phi: A \to B$ *and* $\psi: B \to C$ *be* Λ-*homomorphisms. Show that* $0 \to A \overset{\phi}{\to} B \overset{\psi}{\to} C$ *is exact if and only if the following two conditions are satisfied*:

(a) $\psi\phi = 0$;

(b) *whenever there is a Λ-homomorphism $\theta:X\to B$ such that $\psi\theta=0$,*
then $\theta=\phi f$ for a unique Λ-homomorphism $f:X\to A$.

Solution. Assume that (*a*) and (*b*) both hold. Then $\operatorname{Im}\phi\subseteq\operatorname{Ker}\psi$.
Also if $j:\operatorname{Ker}\psi\to B$ is the inclusion mapping, then $\psi j=0$ and therefore, by (*b*), $j=\phi\gamma$ for some Λ-homomorphism $\gamma:\operatorname{Ker}\psi\to A$. It
follows that $\operatorname{Ker}\psi=\operatorname{Im}j\subseteq\operatorname{Im}\phi$ and therefore $\operatorname{Im}\phi=\operatorname{Ker}\psi$. Now
assume that $f_i:D\to A$ $(i=1,2)$ are Λ-homomorphisms such that
$\phi f_1=\phi f_2=\theta$ (say). Then $\psi\theta=0$ and hence, by (*b*), $f_1=f_2$. It follows,
from Exercise A, that ϕ is an injective mapping. This proves that
$0\to A\overset{\phi}{\to}B\overset{\psi}{\to}C$ is exact. The other assertions contained in the exercise
are trivial.

Exercise D. *Let $\phi:A\to B$ and $\psi:B\to C$ be Λ-homomorphisms. Show*
that $A\overset{\phi}{\to}B\overset{\psi}{\to}C\to0$ is exact if and only if the following two conditions
are satisfied:
(*a*) $\psi\phi=0$;
(*b*) *whenever there is a Λ-homomorphism $\theta:B\to Y$ such that $\theta\phi=0$,*
then $\theta=g\psi$ for a unique Λ-homomorphism $g:C\to Y$.

Solution. Evidently if $A\overset{\phi}{\to}B\overset{\psi}{\to}C\to0$ is exact, then the two conditions are satisfied. Assume therefore that (*a*) and (*b*) hold. Then
$\phi(A)\subseteq\operatorname{Ker}\psi$. Let $\nu:B\to B/\phi(A)$ be the natural mapping. It follows
that $\nu\phi=0$ and therefore, by (*b*), $\nu=\gamma\psi$ for some Λ-homomorphism
$\gamma:C\to B/\phi(A)$. Thus $\phi(A)=\operatorname{Ker}\nu\supseteq\operatorname{Ker}\psi$. Accordingly $\phi(A)=\operatorname{Ker}\psi$
whence $A\overset{\phi}{\to}B\overset{\psi}{\to}C$ is exact.

It remains for us to show that ψ is surjective. Suppose that

$$g_i:C\to D\ (i=1,2)$$

are Λ-homomorphisms such that $g_1\psi=g_2\psi=\theta$ (say). Then $\theta\phi=0$
whence $g_1=g_2$ by condition (*b*). That ψ is surjective now follows from
Exercise B.

2

THE HOM FUNCTOR

2.1 Notation

The notation remains as in section (1.1). In particular Λ, Γ, Δ denote rings (with identity elements) which need not be commutative, and Z denotes the ring of integers. The category of additive abelian groups is identified with the category of Z-modules.

2.2 The Hom functor

Let A and B belong to $\mathscr{C}_\Lambda$. We have already remarked that the Λ-homomorphisms of A into B form an additive abelian group $\operatorname{Hom}_\Lambda (A, B)$. Let $u : A' \to A$ and $v : B \to B'$ be Λ-homomorphisms. We define a mapping

$$\operatorname{Hom}_\Lambda (u, v) : \operatorname{Hom}_\Lambda (A, B) \to \operatorname{Hom}_\Lambda (A', B')$$

by requiring that $f \in \operatorname{Hom}_\Lambda (A, B)$ be mapped into vfu. Clearly $\operatorname{Hom}_\Lambda (u, v)$ is a homomorphism of abelian groups and if u, v are identity maps, then $\operatorname{Hom}_\Lambda (u, v)$ is also an identity map. Again if

$$u' : A'' \to A' \quad \text{and} \quad v' : B' \to B''$$

are in $\mathscr{C}_\Lambda$, then

$$\operatorname{Hom}_\Lambda (uu', v'v) = \operatorname{Hom}_\Lambda (u', v') \operatorname{Hom}_\Lambda (u, v).$$

In fact $\operatorname{Hom}_\Lambda (A, B)$ is a bifunctor from $\mathscr{C}_\Lambda \times \mathscr{C}_\Lambda$ to the category of Z-modules (additive abelian groups) and it is contravariant in the first variable and covariant in the second. This functor is called the *Hom functor*. Note that if $u_1, u_2 : A' \to A$ and $v_1, v_2 : B \to B'$, then

$$\operatorname{Hom}_\Lambda (u_1 + u_2, B) = \operatorname{Hom}_\Lambda (u_1, B) + \operatorname{Hom}_\Lambda (u_2, B)$$

and $\quad \operatorname{Hom}_\Lambda (A, v_1 + v_2) = \operatorname{Hom}_\Lambda (A, v_1) + \operatorname{Hom}_\Lambda (A, v_2).$

Accordingly the Hom functor is additive.

Theorem 1. *The Hom functor is left exact.*

The theorem asserts that if $0 \to A' \to A \to A'' \to 0$ is exact in $\mathscr{C}_\Lambda$, then

$$0 \to \operatorname{Hom}_\Lambda (A'', B) \to \operatorname{Hom}_\Lambda (A, B) \to \operatorname{Hom}_\Lambda (A', B)$$

is exact for every B in $\mathscr{C}_\Lambda$; also if $0 \to B' \to B \to B'' \to 0$ is exact in $\mathscr{C}_\Lambda$, then
$$0 \to \operatorname{Hom}_\Lambda(A, B') \to \operatorname{Hom}_\Lambda(A, B) \to \operatorname{Hom}_\Lambda(A, B'')$$

is exact for every A in $\mathscr{C}_\Lambda$. The verification of these statements is straightforward.

Exercise 1. *Prove Theorem* 1.

Theorem 2. *Let* $\{A_i\}_{i \in I}$ *and* $\{B_j\}_{j \in J}$ *be two families of modules in* $\mathscr{C}_\Lambda$ *and, with the usual notation for direct sums and direct products, put* $A = \bigoplus_i A_i$ *and* $B = \prod_j B_j$. *Then there exists a (canonical) isomorphism* $\operatorname{Hom}_\Lambda(A, B) \approx \prod_{i,j} \operatorname{Hom}_\Lambda(A_i, B_j)$ *of abelian groups.*

Proof. Let $\sigma_i : A_i \to A$ be the canonical injection for the direct sum and $\pi_j : B \to B_j$ the canonical projection for the direct product. If now $f \in \operatorname{Hom}_\Lambda(A, B)$, then $\{\pi_j f \sigma_i\}_{i,j}$ is in $\prod_{i,j} \operatorname{Hom}_\Lambda(A_i, B_j)$. The mapping taking f into $\{\pi_j f \sigma_i\}_{i,j}$ yields the required isomorphism.

Corollary 1. *Suppose that each* A_i *is a* (Λ, Δ)-*bimodule. Then* A *is a* (Λ, Δ)-*bimodule and therefore* $\operatorname{Hom}_\Lambda(A, B)$ *and* $\operatorname{Hom}_\Lambda(A_i, B_j)$ *are* Δ-*modules. On this understanding the canonical isomorphism*
$$\operatorname{Hom}_\Lambda(A, B) \approx \prod_{i,j} \operatorname{Hom}_\Lambda(A_i, B_j)$$
is an isomorphism of Δ-*modules.*

Proof. Suppose that $f \in \operatorname{Hom}_\Lambda(A, B)$ and $\delta \in \Delta$. The product of δ and f is (with a self-explanatory notation) $A \overset{\delta}{\to} A \overset{f}{\to} B$ and the image of this in $\prod_{i,j} \operatorname{Hom}_\Lambda(A_i, B_j)$ has the composite mapping
$$A_i \overset{\sigma_i}{\to} A \overset{\delta}{\to} A \overset{f}{\to} B \overset{\pi_j}{\to} B_j$$
as its typical component. But $A_i \overset{\sigma_i}{\to} A \overset{\delta}{\to} A$ coincides with
$$A_i \overset{\delta}{\to} A_i \overset{\sigma_i}{\to} A.$$

Thus the image of the product of δ and f is the same as the product of δ and the image of f.

Corollary 2. *Suppose that each* B_j *is a* (Λ, Δ)-*bimodule. Then* B *is a* (Λ, Δ)-*bimodule and therefore* $\operatorname{Hom}_\Lambda(A, B)$ *and* $\operatorname{Hom}_\Lambda(A_i, B_j)$ *are* Δ-*modules. On this understanding the canonical isomorphism*
$$\operatorname{Hom}_\Lambda(A, B) \approx \prod_{i,j} \operatorname{Hom}_\Lambda(A_i, B_j)$$
is an isomorphism of Δ-*modules.*

Proof. This is a minor modification of the proof of Corollary 1.

Exercise 2. *Suppose that the modules A, B belong to $\mathscr{C}_\Lambda$ and that Γ is the centre of Λ. Since A is a (Λ, Γ)-bimodule, $\mathrm{Hom}_\Lambda\,(A, B)$ is a Γ-module. Since B is a (Λ, Γ)-bimodule, this also induces a Γ-module structure on $\mathrm{Hom}_\Lambda\,(A, B)$. Show that these two structures coincide.*

Example 1. Let the module A belong to $\mathscr{C}_\Lambda$. Since Λ is a (Λ, Λ)-bimodule, $\mathrm{Hom}_\Lambda\,(\Lambda, A)$ is a Λ-module of the same type as A. We have, in fact, an isomorphism $\mathrm{Hom}_\Lambda\,(\Lambda, A) \xrightarrow{\sim} A$ of Λ-modules in which f is mapped into $f(1)$. Indeed if $A \to A'$ is a Λ-homomorphism, then the diagram

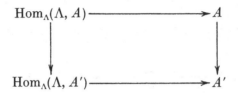

is commutative. Thus $\mathrm{Hom}_\Lambda\,(\Lambda, -)$ is naturally equivalent to the identity functor.

Example 2. Let A belong to $\mathscr{C}_\Lambda$. Since Λ is a (Λ, Λ)-bimodule, $\mathrm{Hom}_\Lambda\,(A, \Lambda)$ is a Λ-module, but of the opposite type to A. Thus $\mathrm{Hom}_\Lambda\,(-, \Lambda)$ is a left exact functor from $\mathscr{C}_\Lambda^L$ to $\mathscr{C}_\Lambda^R$. It is also† a left exact functor from $\mathscr{C}_\Lambda^R$ to $\mathscr{C}_\Lambda^L$. This functor forms the basis of an important duality theory.‡

Example 3. Let Θ be a Z-module and let A belong to $\mathscr{C}_\Lambda^R$. Then A is a (Λ, Z)-bimodule, so $\mathrm{Hom}_Z\,(A, \Theta)$ is in $\mathscr{C}_\Lambda^L$. Again Λ is a (Λ, Λ)-bimodule. Consequently $\mathrm{Hom}_Z\,(\Lambda, \Theta)$ is both a left and a right Λ-module. In fact $\mathrm{Hom}_Z\,(\Lambda, \Theta)$ is a (Λ, Λ)-bimodule and therefore $\mathrm{Hom}_\Lambda\,(A, \mathrm{Hom}_Z\,(\Lambda, \Theta))$ belongs to $\mathscr{C}_\Lambda^L$. We shall compare $\mathrm{Hom}_Z\,(A, \Theta)$ with $\mathrm{Hom}_\Lambda\,(A, \mathrm{Hom}_Z\,(\Lambda, \Theta))$.

Let $f \in \mathrm{Hom}_Z\,(A, \Theta)$ and let $a \in A$. Define $\phi_a : \Lambda \to \Theta$ by $\phi_a(\lambda) = f(a\lambda)$. Then $\phi_a \in \mathrm{Hom}_Z\,(\Lambda, \Theta)$, $\phi_{a_1 + a_2} = \phi_{a_1} + \phi_{a_2}$ and $\phi_{a\lambda} = (\phi_a)\,\lambda$. (For

$$(\phi_a \lambda)\,(\lambda') = \phi_a(\lambda\lambda') = f(a\lambda\lambda')$$

and $\phi_{a\lambda}(\lambda') = f(a\lambda\lambda')$ as well.) Next we define $\phi : A \to \mathrm{Hom}_Z\,(\Lambda, \Theta)$ by

$$\phi(a) = \phi_a.$$

† Strictly speaking we are concerned with two different functors but we allow ourselves a certain informality of language.

‡ See Chapter 5.

Then $\phi \in \operatorname{Hom}_\Lambda (A, \operatorname{Hom}_Z (\Lambda, \Theta))$. We now have a mapping

$$\eta_A : \operatorname{Hom}_Z (A, \Theta) \to \operatorname{Hom}_\Lambda (A, \operatorname{Hom}_Z (\Lambda, \Theta))$$

in which $\eta_A(f) = \phi$ and where, for $a \in A$ and $\lambda \in \Lambda$,

$$(\phi(a))(\lambda) = f(a\lambda).$$

It is easy to check that η_A is an isomorphism of abelian groups. Suppose that f corresponds to ϕ and that $\lambda_0 \in \Lambda$. Then $\lambda_0 f$ and $\lambda_0 \phi$ are both defined. In fact $(\lambda_0 f)(a\lambda) = f(a\lambda\lambda_0)$ and

$$((\lambda_0 \phi)(a))(\lambda) = f(a\lambda\lambda_0)$$

as well. It follows that $\lambda_0 f$ corresponds to $\lambda_0 \phi$. *Consequently η_A is an isomorphism in $\mathscr{C}_\Lambda^L$.*

Suppose next that $u : A' \to A$ is a homomorphism in $\mathscr{C}_\Lambda^R$. Then the diagram

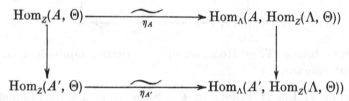

is commutative. These observations may be summed up as follows: *the functors $\operatorname{Hom}_Z (-, \Theta)$ and $\operatorname{Hom}_\Lambda (-, \operatorname{Hom}_Z (\Lambda, \Theta))$ from $\mathscr{C}_\Lambda^R$ to $\mathscr{C}_\Lambda^L$ are naturally equivalent.*

It is, of course, also possible to regard $\operatorname{Hom}_Z (-, \Theta)$ and

$$\operatorname{Hom}_\Lambda (-, \operatorname{Hom}_Z (\Lambda, \Theta))$$

as functors from $\mathscr{C}_\Lambda^L$ to $\mathscr{C}_\Lambda^R$ in which case they will again be naturally equivalent. We shall find a useful application of these equivalences in section (2.5).

2.3 Projective modules

Let P belong to $\mathscr{C}_\Lambda$.

Definition. *P is called a 'projective Λ-module' if the functor*

$$\operatorname{Hom}_\Lambda (P, -),$$

from $\mathscr{C}_\Lambda$ to the category of abelian groups, is exact.

Obviously if a module is projective, then so is any module which is isomorphic to it. The following lemma gives a useful criterion for a module to be projective.

Lemma 1. *Let P belong to $\mathscr{C}_\Lambda$. Then the following two statements are equivalent:*
(a) *P is Λ-projective;*
(b) *whenever we have a diagram*

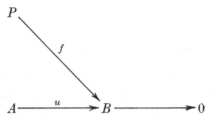

in $\mathscr{C}_\Lambda$ and the row is exact, then there exists a Λ-homomorphism $g\colon P \to A$ such that $f = ug$.

Proof. Condition (b) is equivalent to the following: whenever $A \to B$ is an epimorphism in $\mathscr{C}_\Lambda$, the induced homomorphism

$$\mathrm{Hom}_\Lambda\,(P, A) \to \mathrm{Hom}_\Lambda\,(P, B)$$

is also an epimorphism. In other words, (b) is equivalent to the assumption that $\mathrm{Hom}_\Lambda\,(P, -)$ preserves epimorphisms. The lemma now follows because, in any case, $\mathrm{Hom}_\Lambda\,(P, -)$ is left exact.

We recall that a Λ-module is called *free* if it is isomorphic to a direct sum of copies of Λ or (equivalently) if it has a *base*, that is a linearly independent system of generators.

Theorem 3. *Every free Λ-module is Λ-projective.*
This is a simple application of Lemma 1.

Exercise 3. *Prove Theorem 3.*

Theorem 4. *Let $\{A_i\}_{i \in I}$ be an arbitrary family of Λ-modules and put $A = \bigoplus_i A_i$. Then A is Λ-projective if and only if each A_i is Λ-projective.*
This is another simple application of Lemma 1.

Exercise 4. *Prove Theorem 4.*

Corollary. *Every direct summand of a Λ-projective module is Λ-projective.*

Theorem 5. *If $0 \to A \to B \to P \to 0$ is an exact sequence of Λ-modules and P is Λ-projective, then the sequence splits.†*

Proof. Using Lemma 1, we see that there exists a Λ-homomorphism

† See section (1.3) for the definition of a split exact sequence.

$P \rightarrow B$ such that the composite homomorphism $P \rightarrow B \rightarrow P$ is the identity map of P. The theorem now follows from (Chapter 1, Theorem 4).

Theorem 6. *Let A belong to $\mathscr{C}_\Lambda$. Then there exists an exact sequence $P \rightarrow A \rightarrow 0$, in $\mathscr{C}_\Lambda$, where P is Λ-projective.*

Remark. Note that a canonical construction is given and that, for this construction, P turns out to be *free*.

Proof. Put $F = \underset{\alpha \in A}{\oplus} \Lambda$. Then F is free. Further the mapping $F \rightarrow A$ in which $\{\lambda_\alpha\}_{\alpha \in A}$ is mapped into $\underset{\alpha \in A}{\sum} \lambda_\alpha \alpha$ is a Λ-epimorphism.

Theorem 7. *A module P is projective if and only if it is a direct summand of a free module.*

Proof. If P is a direct summand of a free module, then it is projective by Theorem 3 and Theorem 4 Cor. Suppose now that P is projective. By Theorem 6 we can construct an exact sequence $0 \rightarrow B \rightarrow F \rightarrow P \rightarrow 0$, where F is free. This splits by virtue of Theorem 5. Hence P is (essentially) a direct summand of F.

Corollary. *If the projective module P can be generated by m elements, then P is a direct summand of a free module with a base of m elements.*

Proof. In the latter part of the proof of Theorem 7 we can arrange that F has a base of m elements.

Exercise 5. *Suppose that*

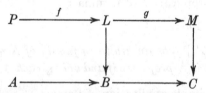

is a commutative diagram in $\mathscr{C}_\Lambda$, that P is projective, $gf = 0$, and the lower row is exact. Deduce that there exists a Λ-homomorphism $P \rightarrow A$ which makes the diagram

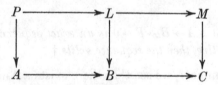

commutative.

Exercise 6.* †*Show that every submodule of a free Z-module is free and deduce that every projective Z-module is free.*

Exercise 7. *Give an example of a projective module which is not free.*

Exercise 8.* *Let P be a projective Λ-module. Show that there exists a free Λ-module F such that $P \oplus F$ and F are isomorphic.*

Theorem 8 (Dual Basis Theorem). *A Λ-module P is projective if and only if there exist families $\{a_i\}_{i \in I}$ of elements of P and $\{f_i\}_{i \in I}$ of elements of $\mathrm{Hom}_\Lambda (P, \Lambda)$ such that (i) for each $a \in P$, $f_i(a) = 0$ for almost all i, and (ii) $a = \Sigma f_i(a) a_i$ for every $a \in P$. When P is projective, the family $\{a_i\}_{i \in I}$ may be taken to be any generating system of P.*

Proof. Assume for the moment that $\psi : F \to P$ is an epimorphism, where F is a free Λ-module with a base $\{e_i\}_{i \in I}$. From (Chapter 1, Lemma 1) and Lemma 1 we see that P is projective if and only if $\psi\phi$ is an identity map for some Λ-homomorphism $\phi : P \to F$.

Suppose first that P is projective and choose such an F, ψ and ϕ. Put $a_i = \psi(e_i)$. If $a \in P$, then we can write $\phi(a) = \Sigma f_i(a) e_i$, where $f_i(a) = 0$ for almost all i and each $f_i : P \to \Lambda$ is a Λ-homomorphism. Further $a = \psi\phi(a) = \Sigma f_i(a) a_i$. Evidently in choosing F and ψ we can arrange that $\{a_i\}_{i \in I}$ is any prescribed generating set of P.

Now suppose that we have families $\{a_i\}_{i \in I}$ and $\{f_i\}_{i \in I}$ satisfying conditions (i) and (ii) in the statement of the theorem. Construct a Λ-homomorphism $\psi : F \to P$ in which F is a free Λ-module with a base $\{e_i\}_{i \in I}$ and $\psi(e_i) = a_i$. Define a Λ-homomorphism $\phi : P \to F$ so that $\phi(a) = \Sigma f_i(a) e_i$. Then $\psi\phi(a) = \sum_i f_i(a) a_i = a$ and therefore $\psi\phi$ is an identity map. Accordingly ψ is epic and now P is projective by our initial remark.

Exercise 9. *Let R be an integral domain with quotient field Q and let $I \neq (0)$ be an ideal of R. Show that the following statements are equivalent:*
  (a) I is projective;
  (b) I is invertible, that is there exist elements $a_1, a_2, \ldots, a_n$ in I and $q_1, q_2, \ldots, q_n$ in Q such that $q_\nu I \subseteq R$ for $\nu = 1, 2, \ldots, n$ and
$$a_1 q_1 + a_2 q_2 + \ldots + a_n q_n = 1.$$
Deduce that a projective ideal of R is necessarily finitely generated.

If R is an integral domain and all its non-zero ideals are invertible, then R is called a *Dedekind ring*.

---

† Exercises marked with an asterisk are likely to prove more difficult than those which are not.

2.4 Injective modules

Let E belong to $\mathscr{C}_\Lambda$.

Definition. *E is said to be an 'injective Λ-module' if* $\mathrm{Hom}_\Lambda\,(-, E)$ *is an exact functor from* $\mathscr{C}_\Lambda$ *to the category of abelian groups.*

Clearly every Λ-module which is isomorphic to an injective module is also injective.

Lemma 2. *Let E belong to $\mathscr{C}_\Lambda$. Then the following two statements are equivalent*:

(a) *E is Λ-injective*;

(b) *whenever we have a diagram*

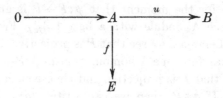

in $\mathscr{C}_\Lambda$ in which the row is exact, there exists a Λ-homomorphism $g: B \to E$ such that $gu = f$.

The proof is similar to that of Lemma 1 and will be omitted.

Theorem 9. *Let $\{B_i\}_{i \in I}$ be a family of Λ-modules and put $B = \prod\limits_{i \in I} B_i$ (direct product). Then B is Λ-injective if and only if each B_i is Λ-injective.*

This is a simple application of Lemma 2.

Corollary. *A direct summand of a Λ-injective module is Λ-injective.*

Exercise 10. *Prove Theorem 9.*

Theorem 10. *Let $0 \to E \overset{f}{\to} A \overset{g}{\to} B \to 0$ be an exact sequence of Λ-modules and suppose that E is injective. Then the sequence splits.*

Proof. By Lemma 2, there exists a Λ-homomorphism $h: A \to E$ such that the diagram

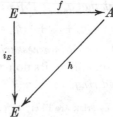

is commutative. That the sequence splits now follows from (Chapter 1, Theorem 4).

Theorem 10 shows that whenever an injective module E is a submodule of a module A, then E is a direct summand of A. We shall see later (Theorem 15) that this is a characteristic property of injective modules.

We can greatly strengthen Lemma 2.

Theorem 11. *Let E be a left Λ-module. Then the following statements are equivalent:*

(a) *E is Λ-injective;*

(b) *for every left ideal I and Λ-homomorphism $f : I \to E$ there exists a Λ-homomorphism $g : \Lambda \to E$ such that the diagram*

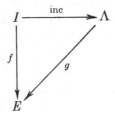

is commutative.

Remark. A similar result holds for right Λ-modules.

Proof. First (a) implies (b) by Lemma 2. *Assume* (b). Suppose that A is a submodule of a left Λ-module B and that we are given a Λ-homomorphism $\phi : A \to E$. By Lemma 2 it is enough to show that ϕ can be extended to a Λ-homomorphism $B \to E$.

Let Σ consist of all pairs (C, ψ), where C is a Λ-module between A and B and ψ is a homomorphism $C \to E$ extending ϕ. Clearly $(A, \phi) \in \Sigma$. If (C_1, ψ_1), (C_2, ψ_2) belong to Σ, let us write $(C_1, \psi_1) \leqslant (C_2, \psi_2)$ if $C_1 \subseteq C_2$ and ψ_2 extends ψ_1. Σ is now partially ordered and it is easy to verify that Σ is a non-empty inductive system. By Zorn's Lemma it has a maximal member (C^*, ψ^*) say. The construction ensures that $A \subseteq C^* \subseteq B$. The proof will be complete if we show that $B = C^*$.

Assume $B \neq C^*$. Choose $\beta \in B$ so that $\beta \notin C^*$. Put

$$I = \{\lambda \,|\, \lambda \in \Lambda, \lambda\beta \in C^*\}.$$

Then I is a left ideal. If $\lambda \in I$, then we can define $f : I \to E$ by

$$f(\lambda) = \psi^*(\lambda\beta).$$

The mapping f is a Λ-homomorphism and, by hypothesis, it can be extended to a Λ-homomorphism $g:\Lambda \to E$. Thus $g(\lambda) = \psi^*(\lambda\beta)$ if $\lambda \in I$.

Put $C = C^* + \Lambda\beta$. If $c \in C$, then we can write $c = c^* + \lambda\beta$, where $c^* \in C^*$ and $\lambda \in \Lambda$. Let $c = c_1^* + \lambda_1\beta$ be a second such representation. Then $\lambda - \lambda_1 \in I$ and therefore $\psi^*(c^*) + g(\lambda) = \psi^*(c_1^*) + g(\lambda_1)$. We can therefore define $\psi:C \to E$ by $\psi(c^* + \lambda\beta) = \psi^*(c^*) + g(\lambda)$ and this will be a Λ-homomorphism. If $a \in A$, then

$$\psi(a) = \psi(a + 0\beta) = \psi^*(a) = f(a).$$

Thus $(C, \psi) \in \Sigma$. Since $C^* \subset C$ and ψ extends ψ^* this contradicts the maximality of (C^*, ψ^*). The proof is now complete.

Example 4. Let R be an integral domain and Q its quotient field. Then Q *is an injective R-module.* For let $f:I \to Q$ be an R-homomorphism, where I is an ideal, then, by Theorem 11, our claim will be established if we can show that f can be extended to an R-homomorphism $\phi:R \to Q$. In this connection we may suppose that $I \neq (0)$. Suppose now that $a, b \in I$ and $a \neq 0$, $b \neq 0$. Since f is an R-homomorphism, it follows that $f(a)/a = f(b)/b$. Thus $f(a)/a = q$ (say) is independent of the choice of a. Define $\phi:R \to Q$ by $\phi(r) = rq$. Then ϕ is an R-homomorphism extending f.

Exercise 11. *Show that every left Λ-module is injective if and only if every left ideal is a direct summand of Λ.*

If Λ has the property that every left ideal is a direct summand of Λ, then Λ is called a *left semi-simple ring*. Right semi-simple rings are defined similarly. It is known that a ring which is left semi-simple is also right semi-simple and vice versa.

We recall that a Λ-module M $\neq$ 0 is said to be *simple* if M and 0 are its only submodules.

Exercise 12.* *Let M be a Λ-module. Show that the following three statements are equivalent:*

   (1) *every submodule of M is a direct summand of M;*
   (2) *M is a sum of simple submodules;*
   (3) *M is a direct sum of simple submodules.*

2.5 Injective Z-modules

It is convenient to make the following:

Definition. *The Z-module Θ is called 'divisible' if $\Theta = m\Theta$ for every non-zero integer m.*

Theorem 12. *A Z-module is injective if and only if it is divisible.*

This follows without difficulty from Theorem 11 provided we note that if $m \in Z$, then mZ is a typical Z-ideal.

Exercise 13. *Prove Theorem 12.*

Theorem 13. *Let P be a projective Λ-module and Θ an injective Z-module. Then $\mathrm{Hom}_Z(P, \Theta)$ is an injective Λ-module of the opposite type to P.*

Proof. Let us first show that $\mathrm{Hom}_Z(\Lambda, \Theta)$ is Λ-injective, i.e. that the functor $\mathrm{Hom}_\Lambda(-, \mathrm{Hom}_Z(\Lambda, \Theta))$ is exact. Now, by Example 3, this functor is naturally equivalent to $\mathrm{Hom}_Z(-, \Theta)$ and this is certainly exact, so the desired conclusion follows.

Next let F be Λ-free. We then have a Λ-isomorphism of the form $F \approx \underset{i \in I}{\oplus} \Lambda$. Thus $\mathrm{Hom}_Z(F, \Theta)$ and $\mathrm{Hom}_Z(\underset{i}{\oplus} \Lambda, \Theta)$ are Λ-isomorphic. It follows, by Theorem 2 Cor. 1, that $\mathrm{Hom}_Z(F, \Theta)$ and the direct product $\underset{i \in I}{\prod} \mathrm{Hom}_Z(\Lambda, \Theta)$ are also Λ-isomorphic. That $\mathrm{Hom}_Z(F, \Theta)$ is Λ-injective now follows from the first part of the proof and Theorem 9.

Finally assume that P is Λ-projective. Then, by Theorem 7, P is a direct summand of a free Λ-module F. Hence $\mathrm{Hom}_Z(P, \Theta)$ is a direct summand of the Λ-injective module $\mathrm{Hom}_Z(F, \Theta)$. That $\mathrm{Hom}_Z(P, \Theta)$ is Λ-injective follows from Theorem 9 Cor. and with this the proof is complete.

Let Q be the additive group of the rational numbers. Clearly Q is divisible. Put $\Omega = Q/Z$. The divisibility of Q is inherited by Ω; consequently (Theorem 12) Ω is Z-injective. For each A in $\mathscr{C}_\Lambda$, put

$$A^* = \mathrm{Hom}_Z(A, \Omega). \tag{2.5.1}$$

Then A^* is a Λ-module of the opposite type to A. Also, by Theorem 13, $\mathrm{Hom}_Z(-, \Omega)$ is a contravariant *exact* functor from $\mathscr{C}_\Lambda^L$ to $\mathscr{C}_\Lambda^R$ and it converts projectives into injectives. Likewise it may be regarded as a contravariant exact functor from $\mathscr{C}_\Lambda^R$ to $\mathscr{C}_\Lambda^L$ again converting projective modules into injective modules.

Lemma 3. *Let X be a Z-module, and x a non-zero element of X. Then there exists $f \in \mathrm{Hom}_Z(X, \Omega)$ such that $f(x) \neq 0$.*

Proof. Since Ω is Z-injective, we may assume that $X = Zx$. First suppose that the period of x is infinite. Then X is a free Z-module with base x and we may choose f so that $f(x)$ is any prescribed non-zero element of Ω. Now suppose that x has period m, Choose $\omega \in \Omega$

so that $\omega \neq 0$ but $m\omega = 0$. (For example ω could be taken to be $(1/m) + Z$.) We now define $f \in \mathrm{Hom}_Z(X, \Omega)$ by $f(nx) = n\omega$. This has the required properties.

As a consequence of Lemma 3 we note that if A is in $\mathscr{C}_\Lambda$, then

$$A^* = 0 \quad \text{if and only if} \quad A = 0. \tag{2.5.2}$$

Theorem 14. *Let A belong to $\mathscr{C}_\Lambda$. Then in $\mathscr{C}_\Lambda$ we can construct an exact sequence $0 \to A \to E$, where E is Λ-injective.*

Remark. It should be noted that the construction can be done in a canonical manner. The theorem asserts that every module is a submodule of an injective module.

Proof. We shall take $\mathscr{C}_\Lambda$ to be $\mathscr{C}_\Lambda^L$. Then A^* belongs to $\mathscr{C}_\Lambda^R$ and we can construct, in a canonical manner (see Theorem 6), an exact sequence $F \to A^* \to 0$, where F is a free Λ-module. Applying the functor $\mathrm{Hom}_Z(-, \Omega)$ to this sequence we obtain an exact sequence $0 \to A^{**} \to F^*$ in $\mathscr{C}_\Lambda^L$ and, by Theorem 13, F^* is Λ-injective.

Let $a \in A$. Define $\phi_a : A^* \to \Omega$ by $\phi_a(f) = f(a)$ for all $f \in A^*$, i.e. for all f in $\mathrm{Hom}_Z(A, \Omega)$. Clearly $\phi_a \in \mathrm{Hom}_Z(A^*, \Omega) = A^{**}$. We therefore have a mapping $\phi : A \to A^{**}$ in which $\phi(a) = \phi_a$. Next ϕ is additive since $\phi_{a_1 + a_2} = \phi_{a_1} + \phi_{a_2}$. If $\lambda \in \Lambda$, then $(\lambda\phi_a)(f) = \phi_a(f\lambda) = (f\lambda)(a)$. But $(f\lambda)(a) = f(\lambda a)$. Consequently $(\lambda\phi_a)(f) = f(\lambda a) = \phi_{\lambda a}(f)$ and therefore $\lambda\phi_a = \phi_{\lambda a}$. This shows that $\phi : A \to A^{**}$ is a homomorphism of left Λ-modules. *It is in fact a monomorphism.* For suppose that $a \in A$, $a \neq 0$. By Lemma 3, there exists $f \in \mathrm{Hom}_Z(A, \Omega) = A^*$ such that $f(a) \neq 0$. It follows that $\phi_a(f) \neq 0$ and hence $\phi_a \neq 0$. This establishes our claim. By combining the monomorphisms $A^{**} \to F^*$ and $A \to A^{**}$, we obtain an exact sequence $0 \to A \to F^*$ in $\mathscr{C}_\Lambda$. With this the theorem is proved.

Theorem 15. *Let E be a Λ-module. Then the following two statements are equivalent:*

    (a) E is Λ-injective;

    (b) E is a direct summand of every Λ-module which contains E as a submodule.

Proof. *Assume* (a). Let V be a Λ-module which contains E as a submodule. By Theorem 10, the exact sequence $0 \to E \to V \to V/E \to 0$ splits and therefore E is a direct summand of V.

Assume (b). By Theorem 14, it is possible to construct an exact sequence $0 \to E \to E^*$, where E^* is Λ-injective. Accordingly E is a submodule of an injective module and hence a direct summand of an injective module by virtue of (b). That E is injective now follows from Theorem 9 Cor.

Exercise 14. *Show that every left Λ-module is projective if and only if every left Λ-module is injective.*

2.6 Essential extensions and injective envelopes

Let A be a submodule of a Λ-module B, i.e. let B be an *extension* of A.

Definition. *B is said to be an 'essential extension' of A if every non-zero submodule of B has a non-zero intersection with A.*

For example, a module is always an essential extension of itself. An extension of a Λ-module A which is different from A will be said to be *non-trivial*.

Lemma 4. *Let A be a submodule of a Λ-module B and let $\{C_i\}_{i\in I}$ be a non-empty family of submodules of B such that if $i_1, i_2 \in I$, then either $C_{i_1} \subseteq C_{i_2}$ or $C_{i_2} \subseteq C_{i_1}$. Denote by C the set-theoretic union of the C_i. Then C is a submodule of B. Further if each C_i is an essential extension of A, then C is an essential extension of A. Finally if $C_i \cap A = 0$ for all i, then $C \cap A = 0$.*

Proof. Only the statement about essential extensions needs comment. Assume therefore that each C_i is an essential extension of A and let U be a non-zero submodule of C. We need only show that U has a non-zero intersection with A and for this we may suppose that U is generated by a single element. But in that case there will exist $i \in I$ such that $U \subseteq C_i$, and now the fact that U has a non-zero intersection with A is clear.

Theorem 16. *Let V be a Λ-module. Then the following two statements are equivalent*:
  (a) *V is Λ-injective*;
  (b) *V has no non-trivial essential extensions.*

Proof. Assume (a) and let U be an essential extension of V. By Theorem 15, $U = V + W$, where W is a submodule of U satisfying

$$W \cap V = 0.$$

Thus $W = 0$ and therefore $U = V$.

Assume (b). By Theorem 14, there exists an injective module E containing V as a submodule. Denote by Σ the set of all submodules of E that intersect V only in the zero submodule, and let us partially order Σ by means of the inclusion relation. Σ is not empty (it contains the zero submodule of E) and, by Lemma 4, it is an inductive system. Accordingly, by Zorn's Lemma, Σ contains a maximal member B say. By construction $V \cap B = 0$. *We claim that $E = V + B$.* Observe that once this claim is established we shall know that V is a direct summand of E and hence injective by Theorem 9 Cor.

Assume that $E \neq V + B$. Put $E' = E/B$ and let V' be the image of V in E'. Then V' is strictly contained in E'. Also V' is isomorphic to V and therefore it inherits property (b). Accordingly E' is not an essential extension of V' and so there exists a submodule C, of E, strictly containing B and such that $V' \cap (C/B) = 0$. It follows that $V \cap C = 0$ and therefore $C \in \Sigma$. This however contradicts the maximal property of B and now the theorem is proved.

Definition. *Let A be a Λ-module. An 'injective envelope' of A is an injective, essential extension of A.*

Theorem 17. *Let A be a Λ-module and E an injective extension module of A. Then E contains a submodule which is an injective envelope of A.*

Remark. In view of Theorem 14, this shows that every Λ-module has an injective envelope. We shall see later that injective envelopes are virtually unique.

Proof. Let Σ consist of all the submodules of E that are essential extensions of A and let Σ be partially ordered by inclusion. Then $A \in \Sigma$ and, by Lemma 4, Σ is an inductive system. Zorn's Lemma now shows that Σ contains a maximal member C say. It will suffice to prove that C is injective and, by Theorem 16, this will follow if we show that C has no non-trivial essential extensions.

Let D be an essential extension of C. Since E is injective, the inclusion mapping $C \to E$ can be extended to a Λ-homomorphism $\phi : D \to E$. Now $\operatorname{Ker} \phi \cap C = 0$ and therefore $\operatorname{Ker} \phi = 0$ by the basic property of essential extensions. Accordingly ϕ is monic and therefore $\phi(D)$ is an essential extension of $\phi(C) = C$ and hence an essential extension of A contained in E. Our choice of C now shows that $\phi(D) = C = \phi(C)$ and therefore $C = D$. This completes the proof.

Theorem 18. *Let $f : A \overset{\sim}{\to} A'$ be an isomorphism of Λ-modules, E an injective envelope of A, and E' an injective envelope of A'. Then f can be extended to an isomorphism of E on to E'.*

Proof. Since E' is injective and A is a submodule of E, Lemma 2 shows that there is a Λ-homomorphism $\phi: E \to E'$ extending f. Now $\operatorname{Ker} \phi \cap A = \operatorname{Ker} f = 0$, whence $\operatorname{Ker} \phi = 0$ because E is an essential extension of A. Accordingly ϕ is monic, and therefore $\phi(E)$ is isomorphic to E and hence injective. We have $A' \subseteq \phi(E)$ and, by Theorem 15, E' is a direct sum of $\phi(E)$ and a module B' say. But then $A' \cap B' = 0$ because $\phi(E) \cap B' = 0$. It follows that $B' = 0$ and hence that $\phi(E) = E'$. Thus ϕ is an isomorphism and the proof is complete.

Let E_1 and E_2 be injective envelopes of the same Λ-module A. By Theorem 18, the identity map of A can be extended to an isomorphism of E_1 on to E_2. In this sense A has, *to within isomorphisms*, a unique injective envelope. From now on we shall therefore speak of *the* injective envelope of a module.

Exercise 15. *Let R be an integral domain. Show that the injective envelope of R (considered as an R-module) is its quotient field. Deduce that R, considered as an R-module, is injective if and only if it is a field.*

Let p be a prime number. The rational numbers whose denominators are powers of p form a ring D_p containing Z. We put

$$Z(p^\infty) = D_p/Z$$

and regard this as a Z-module. As an abelian group $Z(p^\infty)$ is isomorphic to the multiplicative group formed by all p^ν-th roots of unity, where ν is a variable positive integer.

Exercise 16. *Let p be a prime number and ν a positive integer. Show that $Z(p^\infty)$ is the injective envelope of a Z-module which is isomorphic to $Z/p^\nu Z$.*

Let Q denote the rational numbers and let p be a prime number. The rational numbers which have denominators prime to p form a ring which is denoted by Z_p.

Exercise 17. *Show that $Z(p^\infty)$ and Q/Z_p are isomorphic Z-modules.*

Exercise 18. *Let $\{A_i\}_{i \in I}$ be a family of modules in $\mathscr{C}_\Lambda$ and, for each $i \in I$, let B_i be an essential extension of A_i. Show that $\underset{i \in I}{\oplus} B_i$ is an essential extension of $\underset{i \in I}{\oplus} A_i$.*

Exercise 19. *Let $A_1, A_2, \ldots, A_n$ be modules in $\mathscr{C}_\Lambda$ and, for $1 \leqslant i \leqslant n$, let E_i be the injective envelope of A_i. Show that $E_1 \oplus E_2 \oplus \ldots \oplus E_n$ is the injective envelope of $A_1 \oplus A_2 \oplus \ldots \oplus A_n$.*

Let I be a two-sided ideal of Λ. Then Λ/I is not only a ring but it may be regarded as a $(\Lambda, \Lambda/I)$-bimodule with Λ operating on the left and Λ/I on the right. Hence $\mathrm{Hom}_\Lambda (\Lambda/I, -)$ may be regarded as a functor from $\mathscr{C}_\Lambda^L$ to $\mathscr{C}_{\Lambda/I}^L$. We make the following

Definition. *A monomorphism* $f : A \to A'$ *in* $\mathscr{C}_\Lambda$ *is called an 'essential monomorphism' if* A' *is an essential extension of* $f(A)$.

Exercise 20.* *Let* I *be a two-sided ideal of* Λ. *Show that* $\mathrm{Hom}_\Lambda (\Lambda/I, -)$, *considered as a covariant functor from* $\mathscr{C}_\Lambda^L$ *to* $\mathscr{C}_{\Lambda/I}^L$, *preserves injectives and essential monomorphisms.*

Solutions to the Exercises on Chapter 2

Exercise 1. *Show that the bifunctor* $\mathrm{Hom}_\Lambda : \mathscr{C}_\Lambda \times \mathscr{C}_\Lambda \to \mathscr{C}_Z$ *is left exact.*

Solution. Let $0 \to B' \overset{\phi}{\to} B \overset{\psi}{\to} B'' \to 0$ be exact in $\mathscr{C}_\Lambda$ and let A belong to $\mathscr{C}_\Lambda$. Suppose that $f_1, f_2 \in \mathrm{Hom}_\Lambda (A, B')$ and let them be such that $\mathrm{Hom}_\Lambda (A, \phi) (f_1) = \mathrm{Hom}_\Lambda (A, \phi) (f_2)$. Then $\phi f_1 = \phi f_2$. Consequently, since ϕ is a monomorphism, $f_1 = f_2$. Thus $\mathrm{Hom}_\Lambda (A, \phi)$ is a monomorphism. Further,

$$\mathrm{Hom}_\Lambda (A, \psi) \, \mathrm{Hom}_\Lambda (A, \phi) = \mathrm{Hom}_\Lambda (A, \psi\phi) = \mathrm{Hom}_\Lambda (A, 0) = 0$$

since $\mathrm{Hom}_\Lambda (A, -)$ is additive. Accordingly

$$\mathrm{Im} \{\mathrm{Hom}_\Lambda (A, \phi)\} \subseteq \mathrm{Ker} \{\mathrm{Hom}_\Lambda (A, \psi)\}.$$

Let $f : A \to B$ belong to $\mathrm{Ker} \{\mathrm{Hom}_\Lambda (A, \psi)\}$, i.e. suppose that $\psi f = 0$. Then $\mathrm{Im} f \subseteq \mathrm{Ker} \psi = \mathrm{Im} \phi$. But ϕ is a monomorphism. It follows that there is a Λ-homomorphism $g : A \to B'$ which makes the diagram

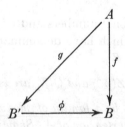

commutative, i.e. which satisfies $\mathrm{Hom}_\Lambda (A, \phi) (g) = f$. Thus

$$\mathrm{Ker} \{\mathrm{Hom}_\Lambda (A, \psi)\} \subseteq \mathrm{Im} \{\mathrm{Hom}_\Lambda (A, \phi)\}$$

and we have shown that

$$0 \to \mathrm{Hom}_\Lambda (A, B') \to \mathrm{Hom}_\Lambda (A, B) \to \mathrm{Hom}_\Lambda (A, B'')$$

is exact.

Let $0 \to A' \xrightarrow{\sigma} A \xrightarrow{\pi} A'' \to 0$ be exact in $\mathscr{C}_\Lambda$ and let B belong to $\mathscr{C}_\Lambda$. If now $g_1, g_2 \in \operatorname{Hom}_\Lambda (A'', B)$ and are such that

$$\operatorname{Hom}_\Lambda (\pi, B)(g_1) = \operatorname{Hom}_\Lambda (\pi, B)(g_2),$$

then $g_1 \pi = g_2 \pi$ and, since π is an epimorphism, $g_1 = g_2$. Accordingly $\operatorname{Hom}_\Lambda (\pi, B)$ is a monomorphism. Next

$$\operatorname{Hom}_\Lambda (\sigma, B)\operatorname{Hom}_\Lambda (\pi, B) = \operatorname{Hom}_\Lambda (\pi\sigma, B) = 0$$

because $\pi\sigma = 0$. Thus we shall have established that

$$0 \to \operatorname{Hom}_\Lambda (A'', B) \to \operatorname{Hom}_\Lambda (A, B) \to \operatorname{Hom}_\Lambda (A', B)$$

is exact and completed the solution if we show that

$$\operatorname{Ker}\{\operatorname{Hom}_\Lambda (\sigma, B)\} \subseteq \operatorname{Im}\{\operatorname{Hom}_\Lambda (\pi, B)\}.$$

Let $f: A \to B$ belong to $\operatorname{Ker}\{\operatorname{Hom}_\Lambda (\sigma, B)\}$, i.e. assume that $f\sigma = 0$. Then $f(\operatorname{Ker}\pi) = f(\operatorname{Im}\sigma) = 0$ and hence there exists a Λ-homomorphism $g: A'' \to B$ which makes

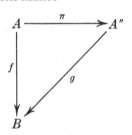

a commutative diagram, i.e. which ensures that $\operatorname{Hom}_\Lambda (\pi, B)(g) = f$. Thus $\operatorname{Ker}\{\operatorname{Hom}_\Lambda (\sigma, B)\} \subseteq \operatorname{Im}\{\operatorname{Hom}_\Lambda (\pi, B)\}$ and the solution is complete.

Exercise 2. *Suppose that modules A, B belong to $\mathscr{C}_\Lambda$ and that Γ is the centre of Λ. Since A is a (Λ, Γ)-bimodule, $\operatorname{Hom}_\Lambda (A, B)$ is a Γ-module. Since B is a (Λ, Γ)-bimodule this also induces a Γ-module structure on $\operatorname{Hom}_\Lambda (A, B)$. Show that these two structures coincide.*

Solution. Let $f \in \operatorname{Hom}_\Lambda (A, B)$ and $\gamma \in \Gamma$. If g is the product of γ and f in the case where A is considered as a (Λ, Γ)-bimodule, then

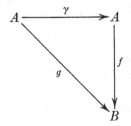

is a commutative diagram. Now let h be the product of γ and f when B is considered as a (Λ, Γ)-bimodule. In this case the diagram

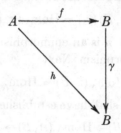

is commutative. However because f is a Λ-homomorphism,

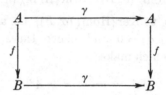

is also commutative. Hence $g = h$ and therefore multiplication by γ is the same in both structures.

Exercise 3. *Show that every free module is projective.*

Solution. Let F be a free Λ-module and consider a diagram

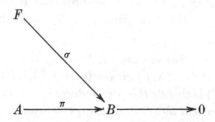

where the row is exact. Let $\{e_i\}_{i \in I}$ be a base for F, and, for each $i \in I$, put $b_i = \sigma(e_i)$. Since π is an epimorphism, for each $i \in I$, there exists $a_i \in A$ such that $\pi(a_i) = b_i$. Define a homomorphism $\tau : F \to A$ by

$$\tau(\sum_{i \in I} \lambda_i e_i) = \sum_{i \in I} \lambda_i a_i.$$

Then $\qquad \pi\tau(\sum_{i \in i} \lambda_i e_i) = \sum_{i \in I} \lambda_i \pi(a_i) = \sum_{i \in I} \lambda_i b_i = \sigma(\sum_{i \in I} \lambda_i e_i).$

Consequently $\pi\tau = \sigma$ and hence F is projective.

Exercise 4. *Let $\{A_i\}_{i \in I}$ be a family of modules in $\mathscr{C}_\Lambda$ and, with the usual notation for direct sums, put $A = \underset{i}{\oplus} A_i$. Show that A is projective if and only if each A_i is projective.*

Solution. Let $\sigma_i : A_i \to A$ and $\pi_i : A \to A_i$ be the canonical homomorphisms. First assume that A is projective and consider a diagram

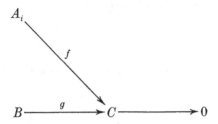

in $\mathscr{C}_\Lambda$, where the row is exact. From consideration of the diagram

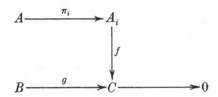

we obtain a homomorphism $\alpha : A \to B$ such that $g\alpha = f\pi_i$. Define $\beta : A_i \to B$ by $\beta = \alpha\sigma_i$. Then $g\beta = g\alpha\sigma_i = f\pi_i\sigma_i = f$. Thus A_i is projective.

Next assume that each A_i is projective and consider a diagram

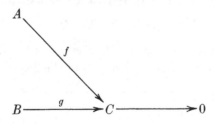

where the row is exact. From the diagram

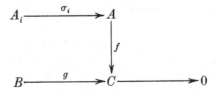

we obtain a homomorphism $\alpha_i : A_i \to B$ such that $g\alpha_i = f\sigma_i$. Define $\beta : A \to B$ by $\beta(a) = \sum_{i \in I} \alpha_i \pi_i(a)$ whenever $a \in A$. This is a Λ-homomorphism. Now, for each $a \in A$,

$$g\beta(a) = \sum_{i \in I} g\alpha_i \pi_i(a) = \sum_{i \in I} f\sigma_i \pi_i(a) = f\left(\sum_{i \in I} \sigma_i \pi_i(a)\right) = f(a),$$

i.e. $g\beta = f$. This shows that A is projective.

Exercise 5. *Suppose that*

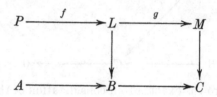

is a commutative diagram in $\mathscr{C}_\Lambda$, where P is Λ-projective, $gf = 0$ and the lower row is exact. Deduce that there is a Λ-homomorphism $P \to A$ which makes the diagram

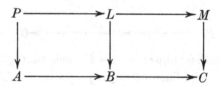

commutative.

Solution. Put $X = \mathrm{Im}\,(A \to B)$. With a self-explanatory notation

$$P \to L \to B \to C = P \to L \to M \to C = 0$$

and therefore $P \to L \to B$ maps P into $\mathrm{Ker}\,(B \to C) = X$. Hence there exists a Λ-homomorphism $P \to X$ such that the diagram

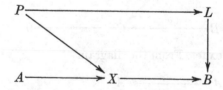

is commutative. But $A \to X$ is an epimorphism and P is projective. It follows that there exists a Λ-homomorphism $P \to A$ which preserves the commutative features of the latter diagram. This mapping has the desired properties.

Exercise 6. *Show that every submodule of a free Z-module is free and deduce that every projective Z-module is free.*

Solution. Let $A \neq 0$ be a free Z-module and B a submodule of A. The module A has a base $\{a_i\}_{i \in I}$ say. We introduce a well ordering $\leqslant$ on I. This secures that I has a minimal element σ say. It is also possible to arrange that it has a maximal element τ.

For $j \in I$ denote by A_j the submodule of A generated by the elements $\{a_i\}_{i \leqslant j}$ and put $B_j = B \cap A_j$. Next we define a Z-homomorphism $f_j : B_j \to Z$ by putting $f_j(\sum_{i \leqslant j} \lambda_i a_i) = \lambda_j$. If now $K_j = \mathrm{Ker} f_j$, then $K_j = B \cap \bigcup_{i < j} A_i$ and we have an exact sequence

$$0 \to K_j \to B_j \to \mathrm{Im} f_j \to 0$$

which splits because, since $\mathrm{Im} f_j$ is an ideal of Z, it is Z-free. We can therefore find a submodule C_j, of B_j, such that $B_j = K_j \oplus C_j$. Here C_j is isomorphic to $\mathrm{Im} f_j$ and so it is Z-free. It will now be proved, using transfinite induction, that
$$B_j = \bigoplus_{i \leqslant j} C_i$$

for all j in I. Since $B_\tau = B \cap A_\tau = B \cap A = B$ this will prove that B is Z-free.

It is clear that $K_\sigma = 0$ and therefore that $B_\sigma = C_\sigma$. Suppose that $q \in I$ and that $B_j = \bigoplus_{i \leqslant j} C_i$ whenever $j < q$. It will suffice to show that $B_q = \bigoplus_{i \leqslant q} C_i$. Now

$$K_q = B \cap (\bigcup_{i < q} A_i) = \bigcup_{i < q} (B \cap A_i) = \bigcup_{i < q} B_i.$$

But $\bigcup_{i < q} B_i$ is the sum of the modules C_i with $i < q$ and this sum is easily seen to be direct. Thus $K_q = \bigoplus_{i < q} C_i$. Finally

$$B_q = K_q \oplus C_q = \bigoplus_{i \leqslant q} C_i$$

and this completes the solution to the first part of the exercise. The second part now follows because a projective Z-module is necessarily a submodule of a free Z-module.

Remark. Let R be an integral domain with the property that every R-ideal can be generated by a single element. The above argument can be adapted to show that any submodule of a free R-module is itself free.

Exercise 7. *Give an example of a projective module which is not free.*

Solution. We have $2Z + 3Z = Z$ and $2Z \cap 3Z = 6Z$. Hence if R is the ring $Z/6Z$, $A = 2Z/6Z$, and $B = 3Z/6Z$, then $R = A \oplus B$ and therefore A is R-projective. But R contains six elements. Consequently a free R-module contains either an infinite number of elements or a finite number k of elements, where k is a multiple of six. However the number of elements in A lies between zero and six. Consequently A is not free.

Exercise 8. *Let P be a projective Λ-module. Show that there exists a free module F such that $P \oplus F$ and F are isomorphic.*

Solution. By Theorem 7, there exists a free Λ-module F' (say) such that $F' = P' \oplus P$ for some Λ-module P'. Put

$$F = P' \oplus P \oplus P' \oplus P \oplus \dots,$$

where there are countably many summands. Then F is free, since it is the direct sum of copies of the free module F'. Furthermore,

$$P \oplus F = P \oplus P' \oplus P \oplus P' \oplus \dots$$
$$= P' \oplus P \oplus P' \oplus P \oplus \dots$$
$$= F.$$

Exercise 9. *Let R be an integral domain with quotient field Q and let $I \neq 0$ be an ideal of R. Show that the following statements are equivalent:*
  (a) *I is projective;*
  (b) *I is invertible, that is there exist elements $a_1, a_2, \dots, a_n$ in I and $q_1, q_2, \dots, q_n$ in Q such that $q_\nu I \subseteq R$ for $\nu = 1, 2, \dots, n$ and $a_1 q_1 + a_2 q_2 + \dots + a_n q_n = 1$.*
Deduce that a projective ideal of R is necessarily finitely generated.

Solution. First assume that I is projective. By Theorem 8, there exist families $\{a_j\}_{j \in J}$ of elements of I and $\{f_j\}_{j \in J}$ of elements of $\mathrm{Hom}_R(I, R)$ such that, for each element x of I, $f_j(x) = 0$ for almost all $j \in J$ and $x = \sum_{j \in J} f_j(x) a_j$. Let a be a non-zero element of I. Then

$$a = f_{j_1}(a) a_{j_1} + f_{j_2}(a) a_{j_2} + \dots + f_{j_n}(a) a_{j_n},$$

where $j_1, j_2, \dots, j_n$ are distinct elements of J. Put $a_\nu = a_{j_\nu}$ and

$$q_\nu = a^{-1} f_{j_\nu}(a)$$

so that $a_1 q_1 + a_2 q_2 + \dots + a_n q_n = 1$. If now $y \in I$, then

$$q_\nu y = a^{-1} f_{j_\nu}(ay) = f_{j_\nu}(y).$$

Thus $q_\nu y \in R$ and we see that I is invertible.

Next assume (*b*) and define a family $\{f_\nu | \nu = 1, 2, ..., n\}$ of elements of $\text{Hom}_R (I, R)$ by $f_\nu(x) = q_\nu x$. (Note that when $x \in I$ we have $q_\nu x \in R$ because $q_\nu I \subseteq R$.) For x in I we now see that

$$x = a_1 q_1 x + a_2 q_2 x + ... + a_n q_n x = a_1 f_1(x) + a_2 f_2(x) + ... + a_n f_n(x).$$

Accordingly, by Theorem 8, I is projective.

Finally suppose that I is a non-zero projective ideal of R and choose $a_1, a_2, ..., a_n$ and $q_1, q_2, ..., q_n$ as above. If now $x \in I$, then

$$x = a_1 q_1 x + a_2 q_2 x + ... + a_n q_n x = a_1 r_1 + a_2 r_2 + ... + a_n r_n,$$

where $r_\nu = q_\nu x \in q_\nu I \subseteq R$. Thus $a_1, a_2, ..., a_n$ generate I and therefore I is finitely generated.

Exercise 10. *Let $\{B_i\}_{i \in I}$ be a family of Λ-modules and let B be their direct product. Show that B is Λ-injective if and only if each B_i is Λ-injective.*

Solution. First assume that B is Λ-injective. Let $i \in I$ and consider a diagram

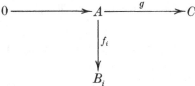

where the row is exact. Let $\sigma_i : B_i \to B$ and $\pi_i : B \to B_i$ be the canonical homomorphisms. Then from the diagram

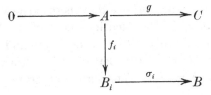

we see there exists a homomorphism $h_i' : C \to B$ such that $h_i' g = \sigma_i f_i$. Define $h_i : C \to B_i$ by $h_i = \pi_i h_i'$. Then $h_i g = \pi_i h_i' g = \pi_i \sigma_i f_i = f_i$. Accordingly B_i is injective.

We next assume that each B_i is injective. Consider a diagram

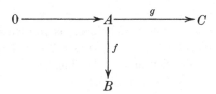

in $\mathscr{C}_\Lambda$, where the row is exact. From the diagrams

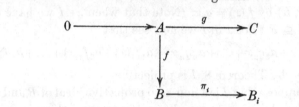

we obtain homomorphisms $k_i: C \to B_i$ such that $\pi_i f = k_i g$. Define a homomorphism $k: C \to B$ by $k(c) = \{k_i(c)\}_{i \in I}$. Then, for each $a \in A$,

$$kg(a) = \{k_i g(a)\}_{i \in I} = \{\pi_i f(a)\}_{i \in I} = f(a).$$

Hence $kg = f$ and B has been shown to be injective.

Exercise 11. *Show that every left Λ-module is injective if and only if each left ideal is a direct summand of Λ.*

Solution. Assume that every left Λ-module is injective. Then every left ideal I is an injective left Λ-module. From the diagram

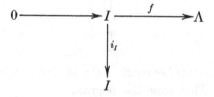

where f is an inclusion map, we see that there exists a homomorphism $g: \Lambda \to I$ such that gf is the identity map of I. Accordingly, by (Chapter 1, Lemma 1), $\mathrm{Im} f = I$ is a direct summand of Λ.

Conversely suppose that every left ideal is a direct summand of Λ and let A be a left Λ-module. Consider the diagram

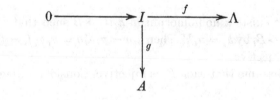

where I is a left ideal of Λ and, as before, f is an inclusion map. Since I is a direct summand of Λ, there exists a homomorphism $\phi: \Lambda \to I$ such that ϕf is the identity map of I. Then $g\phi: \Lambda \to A$ and $g\phi f = g$. Theorem 11 now shows that A is injective.

Exercise 12. *Let M be a Λ-module. Show that the following three statements are equivalent:*

(1) *every submodule of M is a direct summand of M;*
(2) *M is a sum of simple submodules;*
(3) *M is a direct sum of simple submodules.*

Solution. Assume (1). Let N be a submodule of M and U a submodule of N. Then $U \oplus V = M$ for a suitable module V and therefore

$$N = U \oplus (V \cap N).$$

Thus N also has property (1).

Now suppose that $N \neq 0$. *We claim that N contains a simple submodule.* For let x be a non-zero element of N. Using Zorn's Lemma it is easy to see that there exists a maximal submodule A, of N, such that $x \notin A$. By our opening remark, $N = A \oplus B$ for a suitable module B. Evidently $B \neq 0$. Let B' be a non-zero submodule of B. Since B is a submodule of M, we have $B = B' \oplus B''$ for a certain module B''. Since $A \subset A + B' \subseteq N$ we see that $x \in (A + B')$. On the other hand $(A + B') \cap (A + B'') = A$ and therefore $x \notin (A + B'')$. It follows that $B'' \subseteq A$ whence $B'' = 0$ and $B = B'$. This shows that B is simple and establishes our claim.

Let $\{S_j\}_{j \in J}$ be the family consisting of all simple submodules of M. Then $M = (\sum_{j \in J} S_j) \oplus N$ for some submodule N of M. Now

$$(\sum_{j \in J} S_j) \cap N = 0$$

and therefore N cannot contain any simple submodule. It follows that $N = 0$ and therefore $M = \sum_{j \in J} S_j$. Thus (1) implies (2).

Assume (2) and let $\{S_j\}_{j \in J}$ be a family of simple submodules of M satisfying $\sum_{j \in J} S_j = M$. Our intention is to deduce (3) so we may assume that J is non-empty. A straightforward application of Zorn's Lemma shows that there exists a maximal subset I, of J, such that the sum $\sum_{i \in I} S_i$ is direct. Let $j \in J$. Then $S_j \subseteq \sum_{i \in I} S_i$. (For otherwise

$$S_j \cap \sum_{i \in I} S_i = 0$$

and therefore $j \notin I$. Moreover, if $I' = I \cup \{j\}$, then the sum $\sum_{i \in I'} S_i$ is direct and this contradicts the maximality of I.) Accordingly

$$M = \sum_{j \in J} S_j \subseteq \sum_{i \in I} S_i \subseteq M$$

and hence M is the direct sum of the family $\{S_i\}_{i \in I}$. This argument shows that (2) and (3) are equivalent.

Still assuming (2), let N be a submodule of M. Put

$$J_0 = \{j \,|\, j \in J \text{ and } S_j \cap N = 0\}.$$

Then $M = N + \sum\limits_{j \in J_0} S_j$ and under the natural mapping $M \to M/N$ each $S_j \,(j \in J_0)$ is mapped *isomorphically* on to a simple module S_j' say. Thus $M/N = \sum\limits_{j \in J_0} S_j'$ and, by our previous discussion, there exists $I_0 \subseteq J_0$ such that $M/N = \sum\limits_{i \in I_0} S_i'$ and the sum is *direct*. We now have $M = N + (\sum\limits_{i \in I_0} S_i)$. Also, because $\sum\limits_{i \in I_0} S_i'$ is direct, $N \cap (\sum\limits_{i \in I_0} S_i) = 0$. Thus N is a direct summand of M and all is proved.

Exercise 13. *Show that a Z-module is injective if and only if it is divisible.*

Solution. Suppose that E is Z-injective and let m be a non-zero integer. Let $e \in E$ and consider the diagram

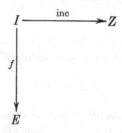

where I is the ideal of Z generated by m and f is the homomorphism defined by $f(nm) = ne$. Since E is injective there is a homomorphism $h : Z \to E$ extending f. Accordingly $e = f(m) = h(m) = mh(1)$. Whence $e \in mE$. It follows that $E = mE$ and the divisibility of E is proved.

Conversely, suppose E is divisible and let I be any non-zero ideal of Z. Consider a diagram

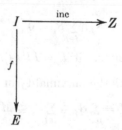

in $\mathscr{C}_Z$. We have $I = Zm$ for some non-zero integer m. Put $e = f(m)$. Since $mE = E$ there is an element e' in E such that $e = me'$. Define a homomorphism $h: Z \to E$ by $h(n) = ne'$. Then h extends f. That E is injective now follows from Theorem 11.

Exercise 14. *Show that every left Λ-module is projective if and only if every left Λ-module is injective.*

Solution. First suppose that every left Λ-module is injective and that A is a left Λ-module. There exists an exact sequence

$$0 \to K \to P \to A \to 0,$$

where P is projective. But K is injective. It follows, from Theorem 10, that the sequence splits. Accordingly A is isomorphic to a direct summand of P and therefore, by the corollary to Theorem 4, A is projective.

Next suppose that every left Λ-module is projective and let B be a left Λ-module. There exists an exact sequence

$$0 \to B \to E \to E/B \to 0,$$

where E is injective. Since E/B is projective, Theorem 5 shows that the sequence splits. Accordingly B is isomorphic to a direct summand of E and therefore B itself is injective.

Exercise 15. *Let R be an integral domain. Show that the injective envelope of R (considered as an R-module) is its quotient field. Deduce that R, considered as an R-module, is injective if and only if R is a field.*

Solution. Let Q be the quotient field of R. Then, by Example 4, Q is injective. Suppose now that A is a non-zero submodule of Q and let r/s be a non-zero member of A, where $r, s \in R$ and $s \neq 0$. Then $r = s(r/s)$ is a non-zero member of $A \cap R$. Hence Q is an essential extension of R and therefore an injective envelope of R.

Finally, R is injective if and only if it is its own injective envelope. Hence R is injective if and only if it is its own quotient field.

Exercise 16. *Let p be a prime number and ν a positive integer. Show that $Z(p^\infty)$ is the injective envelope of a Z-module isomorphic to $Z/p^\nu Z$.*

Solution. We first prove that $Z(p^\infty)$ is divisible and hence, by Theorem 12, injective. Let $\alpha \neq 0$ belong to $Z(p^\infty)$ and let $m \neq 0$ be an integer. Then $\alpha = (a/p^r) + Z$, where $r > 0$, $a \in Z$ and p does not divide a; also m can be expressed in the form $m = p^s n$, where $s \geqslant 0$ and n is an integer not divisible by p. We can now find integers h, k such that $hp^r + kn = a$. Thus $hp^{r+s} + km = ap^s$ and therefore

$$m\left(\frac{k}{p^{r+s}} + Z\right) = \frac{a}{p^r} + Z = \alpha.$$

Accordingly $Z(p^\infty)$ is injective.

Next let C be the Z-submodule of $Z(p^\infty)$ generated by $(1/p^\nu) + Z$. Then C is cyclic with a generator whose period is p^ν and therefore C is isomorphic to $Z/p^\nu Z$. The solution will be complete if we prove that $Z(p^\infty)$ is an essential extension of C and this will follow if we prove that $Z\alpha \cap C \neq 0$. If $r \leqslant \nu$, then α belongs to $Z\alpha$ and C so we may assume that $r > \nu$. But then

$$p^{r-\nu}\alpha = \frac{a}{p^\nu} + Z \subseteq Z\alpha \cap C$$

and $p^{r-\nu}\alpha \neq 0$.

Exercise 17. *Show that $Z(p^\infty)$ and Q/Z_p are isomorphic Z-modules.*

Solution. If, as before, D_p is the Z-module consisting of all rational numbers with denominators which are powers of p, then the inclusion mapping $D_p \to Q$ induces a Z-homomorphism $f: Z(p^\infty) \to Q/Z_p$. Now f is a monomorphism. For if m/p^r, where $m \in Z$ and $r \geqslant 0$, belongs to Z_p, then p^r divides m and therefore m/p^r is in Z.

Now suppose that c and $d \neq 0$ are integers. We can write $d = p^s n$, where $s \geqslant 0$ and n is an integer not divisible by p. Then $c = hp^s + kn$ for suitable integers h and k, and therefore

$$\frac{c}{d} = \frac{k}{p^s} + \frac{h}{n}.$$

Accordingly $\qquad f\left(\frac{k}{p^s} + Z\right) = \frac{k}{p^s} + Z_p = \frac{c}{d} + Z_p.$

This shows that f is an epimorphism as well as a monomorphism and hence it is an isomorphism.

Exercise 18. *Let $\{A_i\}_{i \in I}$ be a family of modules in $\mathscr{C}_\Lambda$, and for each $i \in I$ let B_i be an essential extension of A_i. Show that $\bigoplus_{i \in I} B_i$ is an essential extension of $\bigoplus_{i \in I} A_i$.*

Solution. Let C be a non-zero submodule of $\bigoplus_{i \in I} B_i$. Let $c \in C$ and suppose that $c \neq 0$. Then $c = b_1 + b_2 + \ldots + b_n$. where each b_k is a non-zero member of B_{i_k} and $i_1, i_2, \ldots, i_n$ are distinct elements of I. We shall

construct, in succession, a sequence $\lambda_1, \lambda_2, ..., \lambda_n$ such that
$$\lambda_r(b_1 + b_2 + ... + b_r)$$
is a non-zero member of $A_{i_1} \oplus A_{i_2} \oplus ... \oplus A_{i_r}$. This will ensure that $\lambda_n c$ is a non-zero member of $\underset{i \in I}{\oplus} A_i$ and complete the solution.

Since B_{i_1} is an essential extension of A_{i_1}, there exists $\mu_1 \in \Lambda$ such that $\mu_1 b_1$ is a non-zero member of A_{i_1}. Put $\lambda_1 = \mu_1$. Assume next that we have constructed $\lambda_1, \lambda_2, ..., \lambda_r$ for some $r < n$ to meet our requirements. If $\lambda_r b_{r+1} = 0$ we put $\lambda_{r+1} = \lambda_r$. On the other hand, if
$$\lambda_r b_{r+1} \neq 0,$$
then as $B_{i_{r+1}}$ is an essential extension of $A_{i_{r+1}}$ there exists $\mu_{r+1} \in \Lambda$ such that $\mu_{r+1} \lambda_r b_{r+1}$ is a non-zero member of $A_{i_{r+1}}$. In this case we put $\lambda_{r+1} = \mu_{r+1} \lambda_r$. In either event $\lambda_{r+1}(b_1 + b_2 + ... + b_{r+1})$ is a non-zero member of $A_{i_1} \oplus A_{i_2} \oplus ... \oplus A_{i_{r+1}}$ and the existence of a sequence with the desired properties has been demonstrated.

Exercise 19. *Let $A_1, A_2, ..., A_n$ be modules in $\mathscr{C}_\Lambda$ and, for $1 \leqslant i \leqslant n$, let E_i be the injective envelope of A_i. Show that $E_1 \oplus E_2 \oplus ... \oplus E_n$ is the injective envelope of $A_1 \oplus A_2 \oplus ... \oplus A_n$.*

Solution. This follows at once from the previous exercise and Theorem 9, since a finite direct sum is also a direct product.

Exercise 20. *Let I be a two-sided ideal of Λ. Show that $\mathrm{Hom}_\Lambda(\Lambda/I, -)$, considered as a covariant functor from $\mathscr{C}_\Lambda^L$ to $\mathscr{C}_{\Lambda/I}^L$, preserves injectives and essential monomorphisms.*

Solution. Let A be a module in $\mathscr{C}_\Lambda^L$. Put $0 :_A I = \{a | a \in A, Ia = 0\}$. Then $0 :_A I$ is a Λ-submodule of A which is annihilated by I and hence it is a Λ/I-module. Suppose that $f \in \mathrm{Hom}_\Lambda(\Lambda/I, A)$ and let $\bar{1}$ be the image of the identity element of Λ in Λ/I. If now $\lambda \in I$, then
$$\lambda f(\bar{1}) = f(\lambda \bar{1}) = f(\bar{0}) = 0$$
and therefore $f(\bar{1})$ is in $0 :_A I$. We now have a mapping of $\mathrm{Hom}_\Lambda(\Lambda/I, A)$ into $0 :_A I$ in which f goes into $f(\bar{1})$. An easy verification shows that this is an isomorphism of Λ/I-modules.

Let $\phi : A \to A'$ be a Λ-homomorphism. Then ϕ induces a mapping of $0 :_A I$ into $0 :_{A'} I$. We now have a diagram

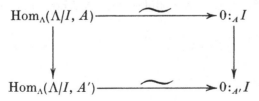

in $\mathscr{C}^L_{\Lambda/I}$, which is commutative and in which the horizontal mappings are isomorphisms.

Let $\phi: A \to A'$ be an essential Λ-monomorphism. In order to show that $\mathrm{Hom}_\Lambda(\Lambda/I, A) \to \mathrm{Hom}_\Lambda(\Lambda/I, A')$ is an essential Λ/I-monomorphism it suffices to show that the same is true of $(0:_A I) \to (0:_{A'} I)$. Alternatively it is enough to prove that if A' is an essential extension of A, then the Λ/I-module $0:_{A'} I$ is an essential extension of $0:_A I$. But this is clear because if B is a submodule of $0:_{A'} I$, then

$$B \cap (0:_A I) = B \cap A.$$

Next suppose that A is Λ-injective. To show that $\mathrm{Hom}_\Lambda(\Lambda/I, A)$ is Λ/I-injective it will suffice to prove that $0:_A I$ is Λ/I-injective. Let

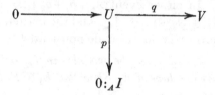

be a diagram in $\mathscr{C}^L_{\Lambda/I}$ with the row exact. Then

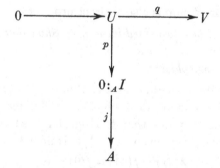

is a diagram in $\mathscr{C}^L_\Lambda$, where j is an inclusion mapping. Since A is Λ-injective, there is a Λ-homomorphism $h: V \to A$ such that $hq = jp$. But $IV = 0$. Consequently $Ih(V) = 0$ and therefore $h(V) \subseteq (0:_A I)$. Thus there exists a Λ-homomorphism $k: V \to (0:_A I)$ such that $jk = h$. It follows that $kq = p$. That $0:_A I$ is Λ/I-injective is now clear because, since V and $0:_A I$ are Λ/I-modules, k is a Λ/I-homomorphism.

3

A DERIVED FUNCTOR

3.1 Notation

The notation remains as in section (1.1). In particular Λ, Γ, Δ denote rings (not necessarily commutative) with identity elements and Z denotes the ring of integers.† As usual we identify the category of additive abelian groups with the category of Z-modules.

3.2 A basic isomorphism

Let A, B belong to $\mathscr{C}_\Lambda$. By (Chapter 2, Theorem 6) and (Chapter 2, Theorem 14), we can construct, in $\mathscr{C}_\Lambda$, exact sequences

$$0 \to A_1 \to P \to A \to 0 \qquad (3.2.1)$$

and
$$0 \to B \to E \to B_1 \to 0, \qquad (3.2.2)$$

where P is projective and E is injective. Since the functor Hom_Λ is left exact, these give rise to exact sequences

$$0 \to \mathrm{Hom}_\Lambda(A, B) \to \mathrm{Hom}_\Lambda(P, B) \overset{\alpha}{\to} \mathrm{Hom}_\Lambda(A_1, B) \to \mathrm{Coker}\,\alpha \to 0$$
$$(3.2.3)$$

and
$$0 \to \mathrm{Hom}_\Lambda(A, B) \to \mathrm{Hom}_\Lambda(A, E) \overset{\beta}{\to} \mathrm{Hom}_\Lambda(A, B_1) \to \mathrm{Coker}\,\beta \to 0 \quad (3.2.4)$$

respectively. We shall now show that $\mathrm{Coker}\,\alpha$ and $\mathrm{Coker}\,\beta$ are isomorphic abelian groups.

Let us say that f in $\mathrm{Hom}_\Lambda(A_1, B)$ is *associated* with g in $\mathrm{Hom}_\Lambda(A, B_1)$ if there exists $u \in \mathrm{Hom}_\Lambda(P, E)$ such that the diagram

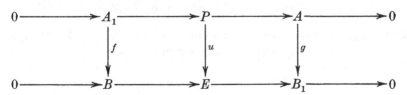

is commutative. Note that given $f: A_1 \to B$ we can always choose u and g to achieve this end. Likewise given g we can find u and f.

† Δ is also used to denote a connecting homomorphism. See section (3.4).

Suppose that $f \in \mathrm{Hom}_\Lambda (A_1, B)$ and that the diagrams

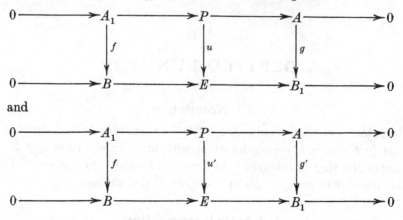

and

are commutative so that g and g' are *both* associated with f. Then

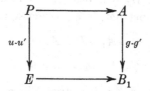

is a commutative diagram and $u - u'$ vanishes on the image of A_1 in P. Hence there exists a Λ-homomorphism $A \to E$ such that

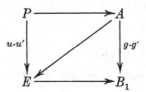

is also commutative, and this shows that

$$g - g' \in \mathrm{Im}\{\mathrm{Hom}_\Lambda (A, E) \xrightarrow{\beta} \mathrm{Hom}_A (A, B_1)\}.$$

Thus if $[g]$ and $[g']$ are the natural images of g and g' in $\mathrm{Coker}\,\beta$, then $[g] = [g']$. In this way we arrive at a Z-homomorphism

$$\mathrm{Hom}_\Lambda (A_1, B) \to \mathrm{Coker}\,\beta \qquad\qquad (3.2.5)$$

in which f is mapped into $[g]$.

Suppose, for the moment, that $f \in \mathrm{Im}\{\mathrm{Hom}_\Lambda (P, B) \xrightarrow{\alpha} \mathrm{Hom}_\Lambda (A_1, B)\}$. Then there exists a Λ-homomorphism $P \to B$ such that the diagram

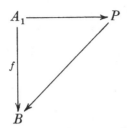

is commutative. Define $u: P \to E$ in $\mathscr{C}_\Lambda$ so that

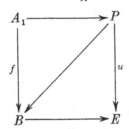

is commutative and now we see that f is associated with the null homomorphism of A into B_1.

The last paragraph shows that the Z-homomorphism (3.2.5) induces a homomorphism

$$\operatorname{Coker}\alpha \to \operatorname{Coker}\beta. \qquad (3.2.6)$$

In this, if f and g are associated in the above sense and $[f]$ and $[g]$ are their natural images in $\operatorname{Coker}\alpha$ and $\operatorname{Coker}\beta$ respectively, then $[f]$ is mapped into $[g]$. Similarly we can define a Z-homomorphism

$$\operatorname{Coker}\beta \to \operatorname{Coker}\alpha \qquad (3.2.7)$$

in which $[g]$ is mapped into $[f]$. The composites of (3.2.6) and (3.2.7) are identity maps. Consequently they are inverse isomorphisms. This yields

Lemma 1. *There is an isomorphism (of abelian groups) between* $\operatorname{Coker}\alpha$ *and* $\operatorname{Coker}\beta$ *with the following property: if* $f \in \operatorname{Hom}_\Lambda(A_1, B)$ *is associated with* $g \in \operatorname{Hom}_\Lambda(A, B_1)$ *in the sense explained above, then the image* $[f]$, *of* f, *in* $\operatorname{Coker}\alpha$ *corresponds to the image* $[g]$, *of* g, *in* $\operatorname{Coker}\beta$.

We shall refer to the isomorphism described in Lemma 1 as the *canonical isomorphism* between $\operatorname{Coker}\alpha$ and $\operatorname{Coker}\beta$.

For the moment let us keep A and B fixed. If we change the sequence (3.2.1) but leave (3.2.2) unaltered, then $\operatorname{Coker}\beta$ will not change. Thus $\operatorname{Coker}\alpha$ will remain unaltered to within isomorphism if we change (3.2.1). Likewise $\operatorname{Coker}\alpha$ is completely unaltered and $\operatorname{Coker}\beta$ is unchanged to within isomorphism if we change (3.2.2). In fact we can say (informally) that $\operatorname{Coker} \alpha = \operatorname{Coker} \beta$ is an abelian group which depends on A and B but is otherwise independent of both (3.2.1) and (3.2.2). We propose to investigate the functorial properties of this group.

Lemma 2. *If either A is projective or B is injective, then $\operatorname{Coker} \alpha$ and $\operatorname{Coker} \beta$ are null.*

Proof. If A is projective, we may take $P = A$ and $A_1 = 0$. Thus $\operatorname{Coker} \alpha = 0$ and therefore $\operatorname{Coker} \beta = 0$ as well. On the other hand, if B is injective, then we may take $E = B$ and $B_1 = 0$ and again the desired conclusion follows.

Let A', A and B, B' belong to $\mathscr{C}_\Lambda$ and let us construct exact sequences

$$0 \to A_1' \to P' \to A' \to 0, \tag{3.2.8}$$

$$0 \to A_1 \to P \to A \to 0, \tag{3.2.9}$$

$$0 \to B \to E \to B_1 \to 0, \tag{3.2.10}$$

and $\qquad 0 \to B' \to E' \to B_1' \to 0, \tag{3.2.11}$

where P', P are projective and E, E' are injective. Suppose now that we are given homomorphisms $\phi: A' \to A$ and $\psi: B \to B'$ in $\mathscr{C}_\Lambda$. We can then construct commutative diagrams

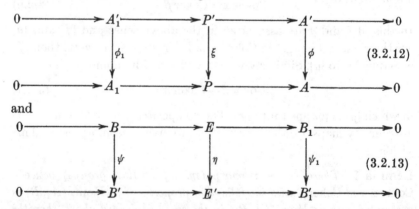

though there may be many ways of doing this. From (3.2.12) we obtain the commutative diagram

whereas (3.2.13) yields

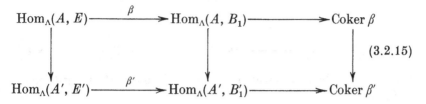

$$\text{(3.2.15)}$$

which is also commutative.

Lemma 3. *The diagram*

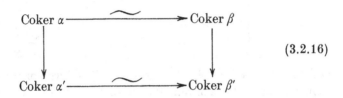

$$\text{(3.2.16)}$$

in which the horizontal mappings are the canonical isomorphisms, is commutative.

Proof. Suppose that the diagram

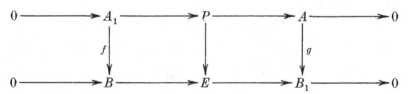

is commutative. Then f and g are associated and therefore the image of f in $\operatorname{Coker}\alpha$ corresponds to the image of g in $\operatorname{Coker}\beta$. Furthermore $\psi f \phi_1$ and $\psi_1 g \phi$ are associated and therefore the image of $\psi f \phi_1$ in $\operatorname{Coker}\alpha'$ corresponds to the image of $\psi_1 g \phi$ in $\operatorname{Coker}\beta'$. This proves the lemma.

Since (3.2.16) *is commutative,* $\operatorname{Coker}\alpha \to \operatorname{Coker}\alpha'$ *is independent of the freedom of choice associated with* ϕ_1 *and* ξ. *Likewise*

$$\operatorname{Coker}\beta \to \operatorname{Coker}\beta'$$

is independent of the freedom of choice associated with η *and* ψ_1.

For example, *any* commutative diagram of the form

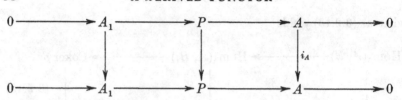

together with the identity map $i_B : B \to B$ will induce the identity map on Coker α.

Suppose that, in $\mathscr{C}_\Lambda$, we have commutative diagrams

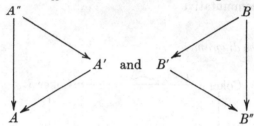

and that for each of A, A', A'' resp. B, B', B'' we construct an exact sequence on the lines of (3.2.9) resp. (3.2.10). In this way we arrive at homomorphisms

$$\text{Coker } \alpha \to \text{Coker } \alpha', \quad \text{Coker } \alpha' \to \text{Coker } \alpha'' \quad \text{and} \quad \text{Coker } \alpha \to \text{Coker } \alpha''$$

(the notation is self-explanatory) and there will be similar mappings involving Coker β, Coker β' and Coker β''. An easy verification shows that the diagram ,

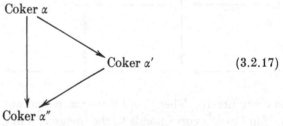

(3.2.17)

is commutative and likewise the same is true of

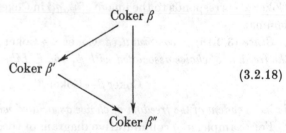

(3.2.18)

It follows from this and the last paragraph, that *if $A' \to A$ and $B \to B'$ are isomorphisms, then* $\operatorname{Coker}\alpha \to \operatorname{Coker}\alpha'$ *and* $\operatorname{Coker}\beta \to \operatorname{Coker}\beta'$ *are isomorphisms as well.*

For each A in $\mathscr{C}_\Lambda$ let us construct an exact sequence

$$0 \to A_1 \to P \to A \to 0 \qquad (3.2.19)$$

with P projective. Once this has been done $\operatorname{Coker}\alpha$ is determined solely by A and B. In fact our results show that $\operatorname{Coker}\alpha$ is a bifunctor from $\mathscr{C}_\Lambda \times \mathscr{C}_\Lambda$ to the category of abelian groups which is contravariant in A and covariant in B. Again, if for each B in $\mathscr{C}_\Lambda$ we construct an exact sequence†

$$0 \to B \to E \to B_1 \to 0, \qquad (3.2.20)$$

then $\operatorname{Coker}\beta$ is likewise a bifunctor from $\mathscr{C}_\Lambda \times \mathscr{C}_\Lambda$ to the category of abelian groups and it is also contravariant in A and covariant in B. Moreover, Lemma 3 shows that the functors $\operatorname{Coker}\alpha$ and $\operatorname{Coker}\beta$ are naturally equivalent.

The functors $\operatorname{Coker}\alpha$ and $\operatorname{Coker}\beta$ depend superficially on the choices of the sequences typified by (3.2.19) and (3.2.20). Suppose now that we make new choices of the sequences of type (3.2.19) but leave those of type (3.2.20) unaltered. Then the functor $\operatorname{Coker}\alpha$ will change but $\operatorname{Coker}\beta$ will not. It follows that, to within equivalence, the functor $\operatorname{Coker}\alpha$ is independent of the sequences (3.2.19) used in its construction and a similar observation applies to $\operatorname{Coker}\beta$.

As this fact is of fundamental importance we shall restate it in a somewhat stronger form. The functor $\operatorname{Coker}\alpha$ resp. $\operatorname{Coker}\beta$ depends initially on the selected sequences (3.2.19) resp. (3.2.20). Suppose that we change these selections quite arbitrarily so that, to typify the change, (3.2.19) becomes

$$0 \to \bar{A}_1 \to \bar{P} \to A \to 0$$

and (3.2.20) $\qquad 0 \to B \to \bar{E} \to \bar{B}_1 \to 0.$

This yields new functors $\operatorname{Coker}\bar{\alpha}$ and $\operatorname{Coker}\bar{\beta}$. By our earlier remarks, the identity maps of A and B yield isomorphisms $\operatorname{Coker}\alpha \xrightarrow{\sim} \operatorname{Coker}\bar{\alpha}$ and $\operatorname{Coker}\beta \xrightarrow{\sim} \operatorname{Coker}\bar{\beta}$. Further, by Lemma 3, the diagram

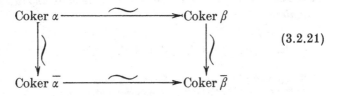

$$(3.2.21)$$

---

† It is understood that E is injective.

is commutative. Indeed if we regard all the terms as functors, then (3.2.21) is a commutative diagram of natural transformations.

The functor that we have just constructed is called the *first extension functor*† and it is denoted by $\mathrm{Ext}^1_\Lambda (A, B)$. More precisely, for each A, B in $\mathscr{C}_\Lambda$ we form a commutative diagram of (abelian) groups

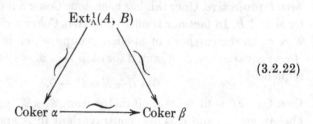

$$(3.2.22)$$

in which all the mappings are isomorphisms and the horizontal one is the canonical isomorphism introduced earlier. To begin with

$$\mathrm{Ext}^1_\Lambda (A, B)$$

is just an abelian group, introduced for convenience, but then we may regard it as a functor (contravariant in A and covariant in B) by requiring that the isomorphisms of (3.2.22) be natural equivalences.‡ In this way we avoid giving preferential treatment to either $\mathrm{Coker}\,\alpha$ or $\mathrm{Coker}\,\beta$.

Let A and B be Λ-modules and let us construct exact sequences $0 \to A_1 \to P \to A \to 0$ and $0 \to B \to E \to B_1 \to 0$, where P is projective and E injective. By the construction of $\mathrm{Ext}^1_\Lambda (A, B)$ we have exact sequences

$$0 \to \mathrm{Hom}_\Lambda (A, B) \to \mathrm{Hom}_\Lambda (P, B) \to \mathrm{Hom}_\Lambda (A_1, B) \to \mathrm{Ext}^1_\Lambda (A, B) \to 0$$

and

$$(3.2.23)$$

$$0 \to \mathrm{Hom}_\Lambda (A, B) \to \mathrm{Hom}_\Lambda (A, E) \to \mathrm{Hom}_\Lambda (A, B_1) \to \mathrm{Ext}^1_\Lambda (A, B) \to 0$$

of Z-modules.

$$(3.2.24)$$

Theorem 1. *Let A be a Λ-module. Then the following two statements are equivalent*:

    (a) *A is projective*;

    (b) $\mathrm{Ext}^1_\Lambda (A, B) = 0$ *for all Λ-modules B*.

---

† The property that gives this functor its name is discussed in section (3.9).

‡ Strictly speaking, Ext^1_Λ represents *two* functors, one for left Λ-modules and one for right Λ-modules. When Λ is commutative, it is not necessary to distinguish between them.

Proof. By Lemma 2, (a) implies (b). We assume (b) and construct an exact sequence $0 \to A_1 \to P \to A \to 0$ with P projective. For each Λ-module B we have an exact sequence (3.2.23). Taking $B = A_1$, and remembering that $\text{Ext}_\Lambda^1 (A, B) = 0$, we find that

$$\text{Hom}_\Lambda (P, A_1) \to \text{Hom}_\Lambda (A_1, A_1) \to 0$$

is exact. Hence there exists a Λ-homomorphism $P \to A_1$ which when composed with $A_1 \to P$ gives an identity map. Accordingly the exact sequence $0 \to A_1 \to P \to A \to 0$ splits and therefore A is isomorphic to a direct summand of P. It follows that A is projective and now the proof is complete.

A similar argument can be used to prove

Theorem 2. *Let B be a Λ-module. Then the following two statements are equivalent:*

(a) B *is injective*;
(b) $\text{Ext}_\Lambda^1 (A, B) = 0$ *for all Λ-modules A.*

Exercise 1. *Prove Theorem 2.*

Theorem 3. *The functor $\text{Ext}_\Lambda^1 (A, B)$ is additive.*

Proof. Let $g: B \to B'$ be a Λ-homomorphism and $0 \to A_1 \to P \to A \to 0$ an exact sequence in $\mathscr{C}_\Lambda$ with P projective. The definition of Ext_Λ^1 as a functor ensures that

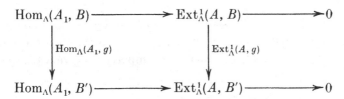

is a commutative diagram with exact rows.

Now suppose that $g_1, g_2: B \to B'$ are Λ-homomorphisms and apply the preceding remark with $g = g_1, g_2$ and $g_1 + g_2$ in turn. Since

$$\text{Hom}_\Lambda (A_1, g_1 + g_2) = \text{Hom}_\Lambda (A_1, g_1) + \text{Hom}_\Lambda (A_1, g_2)$$

it follows that

$$\text{Ext}_\Lambda^1 (A, g_1 + g_2) = \text{Ext}_\Lambda^1 (A, g_1) + \text{Ext}_\Lambda^1 (A, g_2).$$

Thus $\text{Ext}_\Lambda^1 (A, -)$ is additive and a similar argument shows that $\text{Ext}_\Lambda^1 (-, B)$ is additive as well.

Exercise 2. *Suppose that A, B belong to $\mathscr{C}_\Lambda$ and that Γ is the centre of Λ. Since A is a (Λ, Γ)-bimodule, $\operatorname{Ext}_\Lambda^1(A, B)$ is a Γ-module. Since B is a (Λ, Γ)-bimodule, this also induces a Γ-module structure on $\operatorname{Ext}_\Lambda^1(A, B)$. Show that these two structures coincide.*

3.3 Some remarks on diagrams

Let $\{D_i\}_{i \in I}$ be a family of modules in $\mathscr{C}_\Lambda$ and Ω a set of ordered pairs of elements of I. Assume that with each pair $(i, j) \in \Omega$ there is associated a Λ-homomorphism $\phi_{ij} : D_i \to D_j$ and denote by D the complete system $[D_i, \phi_{ij}; i \in I, (i, j) \in \Omega]$. We shall call D a *diagram* in $\mathscr{C}_\Lambda$. Two diagrams

$$D = [D_i, \phi_{ij}; i \in I, (i, j) \in \Omega] \tag{3.3.1}$$

and $$D' = [D_i', \phi_{ij}'; i \in I, (i, j) \in \Omega] \tag{3.3.2}$$

which share the same labels I and Ω will be said to be *similar*.

Let D and D' be similar diagrams, the details being as shown in (3.3.1) and (3.3.2). By a *translation* $f : D \to D'$ will be meant a family $\{f_i\}_{i \in I}$ of Λ-homomorphisms $f_i : D_i \to D_i'$ such that for each (i, j) in Ω the square

$$\begin{array}{ccc} D_i & \xrightarrow{\phi_{ij}} & D_j \\ \downarrow{\scriptstyle f_i} & & \downarrow{\scriptstyle f_j} \\ D_i' & \xrightarrow{\phi_{ij}'} & D_j' \end{array}$$

is commutative. Two translations $f : D \to D'$ and $g : D \to D'$ of diagrams may be added to give a new translation of D into D'. Also translations such as $D \to D'$ and $D' \to D''$ may be composed to give a translation $D \to D''$.

3.4 The Ker–Coker sequence

In section (3.4) it is to be understood that all diagrams are formed in $\mathscr{C}_\Lambda$. Thus all the objects in the diagrams will be Λ-modules and all the mappings Λ-homomorphisms. We note that, given a commutative diagram

$$\begin{array}{ccc} L & \xrightarrow{\phi} & M \\ \downarrow & & \downarrow \\ U & \xrightarrow{\psi} & V \end{array}$$

there exist unique homomorphisms

$$\text{Ker}\,\phi \to \text{Ker}\,\psi \quad \text{and} \quad \text{Coker}\,\phi \to \text{Coker}\,\psi$$

which make the enlarged configuration

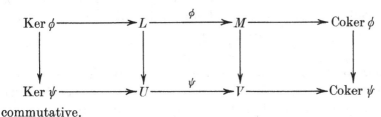

commutative.

Lemma 4. *Let*

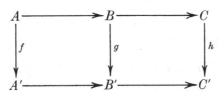

be a commutative diagram with exact rows. If now $A' \to B'$ is monic, then the induced sequence $\text{Ker}\,f \to \text{Ker}\,g \to \text{Ker}\,h$ is exact. On the other hand, if $B \to C$ is epic, then the resulting sequence

$$\text{Coker}\,f \to \text{Coker}\,g \to \text{Coker}\,h$$

is exact.

Proof. Assume that $A' \to B'$ is monic. Obviously the product of $\text{Ker}\,f \to \text{Ker}\,g$ and $\text{Ker}\,g \to \text{Ker}\,h$ is null. Let b in $\text{Ker}\,g$ become zero in $\text{Ker}\,h$. Then b is the image with respect to $A \to B$ of some element $a \in A$. Now $f(a)$ becomes $g(b) = 0$ in B' and therefore $f(a) = 0$, i.e. $a \in \text{Ker}\,f$, because $A' \to B'$ is monic. This proves that

$$\text{Ker}\,f \to \text{Ker}\,g \to \text{Ker}\,h$$

is exact. The other sequence is treated similarly.

Exercise 3. *Complete the proof of Lemma 4.*

Until we come to Theorem 4 we shall be concerned with a commutative diagram

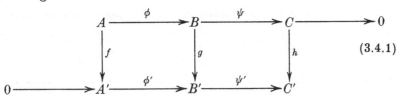

(3.4.1)

with exact rows. We now define a Λ-homomorphism

$$\Delta : \operatorname{Ker} h \to \operatorname{Coker} f. \tag{3.4.2}$$

The construction of Δ. Let $c \in \operatorname{Ker} h \subseteq C$. Since ψ is an epimorphism, $c = \psi(b)$ for some element $b \in B$ and then $\psi' g(b) = h\psi(b) = h(c) = 0$. Thus $g(b) = \phi'(a')$ for some $a' \in A'$ and a' itself has a natural image, $[a']$ say, in $\operatorname{Coker} f$. The mapping Δ is now defined by

$$\Delta(c) = [a']. \tag{3.4.3}$$

In this construction the element b is not unique. However if we change it the effect on a' is to replace it by an element of the form $a' + f(a)$, where $a \in A$. This does not alter $[a']$. Thus Δ is well defined and, after this observation, it is easily seen to be a Λ-homomorphism. It is called the *connecting homomorphism*.

Theorem 4. *Let* (3.4.1) *be a commutative diagram with exact rows. Then the resulting sequence* (*the* Ker–Coker *sequence*)

$$\operatorname{Ker} f \to \operatorname{Ker} g \to \operatorname{Ker} h \overset{\Delta}{\to} \operatorname{Coker} f \to \operatorname{Coker} g \to \operatorname{Coker} h \tag{3.4.4}$$

is exact.

Proof. By Lemma 4, it is enough to show that

$$\operatorname{Ker} g \to \operatorname{Ker} h \overset{\Delta}{\to} \operatorname{Coker} f \tag{3.4.5}$$

and

$$\operatorname{Ker} h \overset{\Delta}{\to} \operatorname{Coker} f \to \operatorname{Coker} g \tag{3.4.6}$$

are both exact. Let $b \in \operatorname{Ker} g$ and consider $\Delta(\psi(b))$. Since

$$g(b) = 0 = \phi'(0),$$

it follows that $\Delta(\psi(b))$ is the image of zero in $\operatorname{Coker} f$. Thus $\Delta(\psi(b)) = 0$ and therefore the result of combining $\operatorname{Ker} g \to \operatorname{Ker} h$ with Δ is null. Now suppose that $c \in \operatorname{Ker} h$ and $\Delta(c) = 0$. Let b and a' be defined as in the construction of $\Delta(c)$. Then $a' = f(a)$ for some $a \in A$. It follows that $g(b) = \phi'(a') = \phi' f(a) = g\phi(a)$ and therefore $b - \phi(a) \in \operatorname{Ker} g$. Thus $c = \psi(b) = \psi(b - \phi(a)) \in \psi (\operatorname{Ker} g)$. This proves that (3.4.5) is exact. The proof that (3.4.6) is exact is left as an exercise.

Exercise 4. *Complete the proof of Theorem* 4.

It is worth noting that in the case where we have a more detailed commutative diagram

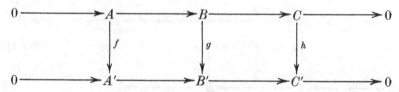

with exact rows, then the Ker–Coker sequence can be extended at both ends to yield an exact sequence

$$0 \to \mathrm{Ker}\, f \to \mathrm{Ker}\, g \to \mathrm{Ker}\, h \xrightarrow{\Delta} \mathrm{Coker}\, f \to \mathrm{Coker}\, g \to \mathrm{Coker}\, h \to 0.$$

Corollary. *Let*

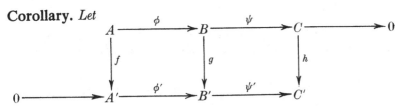

be a commutative diagram with exact rows and let T be a translation of this into a diagram

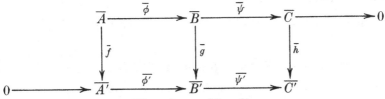

with similar properties. Then the resulting diagram

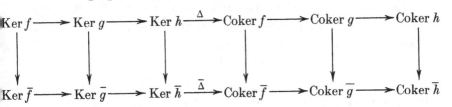

is commutative.

Proof. It is enough to show that

$$
\begin{array}{ccc}
\mathrm{Ker}\, h & \xrightarrow{\ \Delta\ } & \mathrm{Coker}\, f \\
\downarrow & & \downarrow \\
\mathrm{Ker}\, \bar{h} & \xrightarrow{\ \bar{\Delta}\ } & \mathrm{Coker}\, \bar{f}
\end{array}
\tag{3.4.7}
$$

is commutative. Let $c \in \mathrm{Ker}\, h$ and let b and a' be chosen as in the construction of $\Delta(c)$. Then the image of $\Delta(c)$ in $\mathrm{Coker}\,\bar{f}$ may be obtained as the image of $T_{A'}(a')$, in $\mathrm{Coker}\,\bar{f}$, where $T_{A'} \colon A' \to \bar{A}'$ denotes the translation homomorphism. The image of c in $\mathrm{Ker}\,\bar{h}$ is, with a similar

3   N F C

notation, $T_C(c)$. Now to construct $\bar{\Delta}(T_C(c))$ we may use the elements $T_B(b)$ and $T_{A'}(a')$ in the same way that b and a' were used in the construction of $\Delta(c)$. Thus $\bar{\Delta}(T_C(c))$ is simply the image of $T_{A'}(a')$ in $\text{Coker}\,\bar{f}$, i.e. it is the image of $\Delta(c)$ in $\text{Coker}\,\bar{f}$. This proves that (3.4.7) is commutative.

Exercise 5. *Suppose that the sequence $0 \to A \to B \to C \to 0$ is exact in $\mathscr{C}_\Lambda$ and let γ be a central element† of Λ. Establish the existence of an exact sequence*

$$0 \to (0 :_A \gamma) \to (0 :_B \gamma) \to (0 :_C \gamma) \to A/\gamma A \to B/\gamma B \to C/\gamma C \to 0,$$

where, for example, $0 :_A \gamma$ is the submodule of A annihilated by γ.

Exercise 6. *Let $f : A \to B$ and $g : B \to C$ be Λ-homomorphisms. Show that there exists an exact sequence*

$$0 \to \text{Ker}\,f \to \text{Ker}\,(gf) \to \text{Ker}\,g \to \text{Coker}\,f \to \text{Coker}\,(gf) \to \text{Coker}\,g \to 0.$$

3.5 Further properties of Ext_Λ^1

Throughout section (3.5) we shall work in $\mathscr{C}_\Lambda$. Accordingly all mappings are understood to be Λ-homomorphisms of Λ-modules.

Let $0 \to A' \to A \to A'' \to 0$ be exact and let $0 \to B \to E \to B_1 \to 0$ also be exact but with E injective. Then there results a commutative diagram

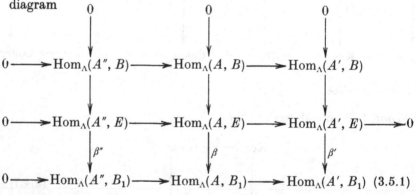

with exact rows and columns. Note that the middle row, which is longer than the others, is exact because E is injective. The theory of the Ker–Coker sequence now yields the exact sequence

$$\text{Ker}\,\beta'' \to \text{Ker}\,\beta \to \text{Ker}\,\beta' \xrightarrow{\Delta} \text{Coker}\,\beta'' \to \text{Coker}\,\beta \to \text{Coker}\,\beta' \quad (3.5.2)$$

† A *central element* of Λ is one which belongs to the centre of the ring.

and this in turn leads to the exact sequence

$$0 \to \mathrm{Hom}_\Lambda (A'', B) \to \mathrm{Hom}_\Lambda (A, B) \to \mathrm{Hom}_\Lambda (A', B)$$

$$\overset{\Delta}{\to} \mathrm{Coker}\, \beta'' \to \mathrm{Coker}\, \beta \to \mathrm{Coker}\, \beta'. \quad (3.5.3)$$

Finally, because of the equivalences (3.2.22), this can be converted into an exact sequence

$$0 \to \mathrm{Hom}_\Lambda (A'', B) \to \mathrm{Hom}_\Lambda (A, B) \to \mathrm{Hom}_\Lambda (A', B)$$

$$\overset{\Delta}{\to} \mathrm{Ext}^1_\Lambda (A'', B) \to \mathrm{Ext}^1_\Lambda (A, B) \to \mathrm{Ext}^1_\Lambda (A', B). \quad (3.5.4)$$

We pause here to make an observation. Suppose that in the above discussion we replace $0 \to A' \to A \to A'' \to 0$ by an exact sequence $0 \to A_1 \to P \to A \to 0$ with P projective. Then the diagram (3.5.1) takes the form

$$
\begin{array}{ccccccc}
0 & & 0 & & 0 & & \\
\downarrow & & \downarrow & & \downarrow & & \\
0 \longrightarrow \mathrm{Hom}_\Lambda(A, B) & \longrightarrow & \mathrm{Hom}_\Lambda(P, B) & \longrightarrow & \mathrm{Hom}_\Lambda(A_1, B) & & \\
\downarrow & & \downarrow & & \downarrow & & \\
0 \longrightarrow \mathrm{Hom}_\Lambda(A, E) & \longrightarrow & \mathrm{Hom}_\Lambda(P, E) & \longrightarrow & \mathrm{Hom}_\Lambda(A_1, E) & \longrightarrow & 0 \\
\downarrow & & \downarrow & & \downarrow & & \\
0 \longrightarrow \mathrm{Hom}_\Lambda(A, B_1) & \longrightarrow & \mathrm{Hom}_\Lambda(P, B_1) & \longrightarrow & \mathrm{Hom}_\Lambda(A_1, B_1) & & (3.5.5)
\end{array}
$$

and the connecting homomorphism in (3.5.3) is now

$$\Delta : \mathrm{Hom}_\Lambda (A_1, B) \to \mathrm{Coker}\, \{\mathrm{Hom}_\Lambda (A, E) \to \mathrm{Hom}_\Lambda (A, B_1)\}.$$

It is not difficult to show that this is the same mapping that we encountered earlier in (3.2.5). Since we shall not make use of this fact, the verification of its correctness is left to the reader.

Theorem 5. *Every exact sequence* $0 \to A' \to A \to A'' \to 0$ *gives rise to an exact sequence*

$$0 \to \mathrm{Hom}_\Lambda (A'', B) \to \mathrm{Hom}_\Lambda (A, B) \to \mathrm{Hom}_\Lambda (A', B)$$

$$\overset{\Delta}{\to} \mathrm{Ext}^1_\Lambda (A'', B) \to \mathrm{Ext}^1_\Lambda (A, B) \to \mathrm{Ext}^1_\Lambda (A', B). \quad (3.5.6)$$

Furthermore if

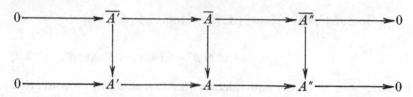

is a commutative diagram in $\mathscr{C}_\Lambda$ with exact rows and $B \to \bar{B}$ is a Λ-homomorphism, then (with a self-explanatory notation) the diagram

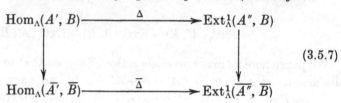

(3.5.7)

is commutative.

Proof. The first statement has already been proved. We keep the notation as before and construct an additional exact sequence

$$0 \to \bar{B} \to \bar{E} \to \bar{B}_1 \to 0,$$

with $\bar{E}$ injective. We can now form a commutative diagram

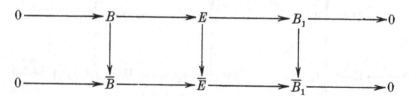

where $B \to \bar{B}$ is the given homomorphism. This together with the commutative diagram

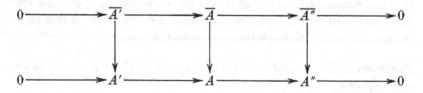

induces a translation of (3.5.1) and now the desired result follows from Theorem 4 Cor.

Theorem 5 has an important consequence to which we must draw attention. Let $0 \to A' \to A \to A'' \to 0$ be an exact sequence of Λ-modules. If now $0 \to B \to E \to B_1 \to 0$ is also exact and E is injective, then we can use these sequences to construct a connecting homomorphism

$$\Delta : \text{Hom}_\Lambda (A', B) \to \text{Ext}_\Lambda^1 (A'', B).$$

Now suppose that in place of $0 \to B \to E \to B_1 \to 0$ we employ a second exact sequence $0 \to B \to \bar{E} \to \bar{B}_1 \to 0$, where $\bar{E}$ is also injective. This time the construction produces a connecting homomorphism

$$\bar{\Delta} : \text{Hom}_\Lambda (A', B) \to \text{Ext}_\Lambda^1 (A'', B)$$

say. However if we apply Theorem 5 with $0 \to \bar{A}' \to \bar{A} \to \bar{A}'' \to 0$ identical to $0 \to A' \to A \to A'' \to 0$ and use $0 \to B \to \bar{E} \to \bar{B}_1 \to 0$ in place of $0 \to \bar{B} \to \bar{E} \to \bar{B}_1 \to 0$ ($B \to \bar{B}$ being, in this case, the identity mapping), then the commutative property of (3.5.7) shows that $\Delta = \bar{\Delta}$. Accordingly *the connecting homomorphism* $\Delta : \text{Hom}_\Lambda (A', B) \to \text{Ext}_\Lambda^1 (A'', B)$ *is independent of the choice of the sequence* $0 \to B \to E \to B_1 \to 0$ *used in its construction.*

We turn next to a companion result to Theorem 5. Let

$$0 \to B' \to B \to B'' \to 0$$

be an exact sequence and suppose that $0 \to A_1 \to P \to A \to 0$ is also exact with P projective. From these we obtain a commutative diagram

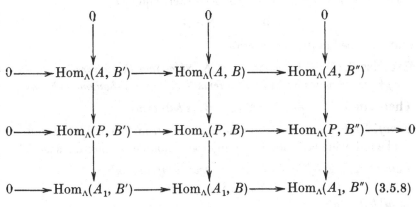

with exact rows and columns, the longer middle row being exact because P is projective.

From this we obtain, by considerations very similar to those already encountered, the theorem which follows.

Theorem 6. *Every exact sequence* $0 \to B' \to B \to B'' \to 0$ *gives rise to an exact sequence*

$$0 \to \mathrm{Hom}_\Lambda (A, B') \to \mathrm{Hom}_\Lambda (A, B) \to \mathrm{Hom}_\Lambda (A, B'')$$
$$\overset{\Delta}{\to} \mathrm{Ext}^1_\Lambda (A, B') \to \mathrm{Ext}^1_\Lambda (A, B) \to \mathrm{Ext}^1_\Lambda (A, B''). \quad (3.5.9)$$

Further if

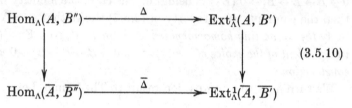

is a commutative diagram in $\mathscr{C}_\Lambda$ *with exact rows and* $\bar{A} \to A$ *is a* Λ-*homomorphism, then (with a self-explanatory notation) the diagram*

$$
\begin{array}{ccc}
\mathrm{Hom}_\Lambda(A, B'') & \overset{\Delta}{\longrightarrow} & \mathrm{Ext}^1_\Lambda(A, B') \\
\downarrow & & \downarrow \\
\mathrm{Hom}_\Lambda(\bar{A}, \bar{B}'') & \overset{\bar{\Delta}}{\longrightarrow} & \mathrm{Ext}^1_\Lambda(\bar{A}, \bar{B}')
\end{array}
\quad (3.5.10)
$$

is commutative.

The reader should note that it follows from this theorem (by virtue of observations analogous to those made after the proof of Theorem 5) that the connecting homomorphism $\Delta : \mathrm{Hom}_\Lambda (A, B'') \to \mathrm{Ext}^1_\Lambda (A, B')$ does not depend on the choice of the exact sequence

$$0 \to A_1 \to P \to A \to 0$$

that was used in its construction.

Exercise 7. *Let* Γ *be the centre of* Λ. *Show that the connecting homomorphisms (denoted by* Δ) *in Theorems 5 and 6 are* Γ-*homomorphisms.*†

Theorem 7. *The functor* $\mathrm{Ext}^1_\Lambda (A, B)$ *is half exact.*

Proof. Half exactness follows immediately from Theorems 5 and 6.

The following criterion for an injective module is sometimes useful.

Theorem 8. *A left* Λ-*module* E *is injective if and only if*

$$\mathrm{Ext}^1_\Lambda (\Lambda/I, E) = 0$$

for all left ideals I.

Proof. We shall assume that $\mathrm{Ext}^1_\Lambda (\Lambda/I, E) = 0$ for all left ideals I and deduce that E is injective. (The converse follows by Theorem 2.)

---

† See (Chapter 2, Exercise 2) and Exercise 2.

Left I be a left ideal and $f: I \to E$ a Λ-homomorphism. By (Chapter 2, Theorem 11) it is enough to show that f can be extended to a homomorphism of Λ into E. Now this will follow if we show that the homomorphism $\text{Hom}_\Lambda(\Lambda, E) \to \text{Hom}_\Lambda(I, E)$ induced by the inclusion mapping $I \to \Lambda$ is surjective. But the sequence $0 \to I \to \Lambda \to \Lambda/I \to 0$ is exact. Consequently, by Theorem 5, we have an exact sequence

$$0 \to \text{Hom}_\Lambda(\Lambda/I, E) \to \text{Hom}_\Lambda(\Lambda, E) \to \text{Hom}_\Lambda(I, E) \to \text{Ext}_\Lambda^1(\Lambda/I, E).$$

However $\text{Ext}_\Lambda^1(\Lambda/I, E) = 0$. Consequently $\text{Hom}_\Lambda(\Lambda, E) \to \text{Hom}_\Lambda(I, E)$ is surjective as required.

We recall that a module is called *cyclic* if it can be generated by a single element.

Exercise 8. *Show that if each cyclic left Λ-module is projective, then all left Λ-modules are projective.*

Theorem 9. *Let $0 \to A' \to A \to A'' \to 0$ be a split exact sequence of Λ-modules. Then, for each B in $\mathscr{C}_\Lambda$, the connecting homomorphism*

$$\Delta: \text{Hom}_\Lambda(A', B) \to \text{Ext}_\Lambda^1(A'', B)$$

is null.

Proof. We have an exact sequence

$$\text{Hom}_\Lambda(A, B) \to \text{Hom}_\Lambda(A', B) \to \text{Ext}_\Lambda^1(A'', B)$$

and, by (Chapter 1, Theorem 9), $\text{Hom}_\Lambda(A, B) \to \text{Hom}_\Lambda(A', B)$ is surjective. The theorem follows. Similarly we can prove

Theorem 10. *Let $0 \to B' \to B \to B'' \to 0$ be a split exact sequence of Λ-modules. Then for each A in $\mathscr{C}_\Lambda$, the connecting homomorphism*

$$\Delta: \text{Hom}_\Lambda(A, B'') \to \text{Ext}_\Lambda^1(A, B')$$

is null.

3.6 Consequences of the vanishing of $\text{Ext}_\Lambda^1(A, B)$

The following theorem will help to throw new light on the significance of the extension functor though we shall postpone consideration of some of its implications until later.

Theorem 11. *Let $0 \to B \overset{g}{\to} X \overset{f}{\to} A \to 0$ be an exact sequence in $\mathscr{C}_\Lambda$ and let $\gamma: B \to B'$ be a Λ-homomorphism. Then it is possible to construct a*

commutative diagram

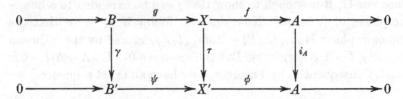

in $\mathscr{C}_\Lambda$, *with exact rows.*

Proof. Let $\mu: B \to B' \oplus X$ be the Λ-homomorphism defined by

$$\mu(b) = (-\gamma(b), g(b)).$$

Then because g is a monomorphism so too is μ and we can construct an exact sequence

$$0 \to B \overset{\mu}{\to} B' \oplus X \overset{\pi}{\to} X' \to 0.$$

Now define $\psi: B' \to X'$ by $\psi(b') = \pi(b', 0)$. If $\psi(b') = 0$, then

$$(b', 0) = (-\gamma(b), g(b))$$

for some b in B and therefore $b = 0$ and hence $b' = 0$ as well. It follows that ψ is a monomorphism. Next if $\pi(b', x) = \pi(b_1', x_1)$, then $x_1 - x = g(\beta)$ for some β in B. Consequently $f(x_1) = f(x)$. We can therefore define a Λ-homomorphism $\phi: X' \to A$ by $\phi\pi(b', x) = f(x)$. Evidently ϕ is an epimorphism and $\phi\psi = 0$. Now suppose that $\phi\pi(b', x) = 0$. Then $x = g(b_0)$ for some $b_0 \in B$ and therefore

$$(b', x) = (b_0', 0) + \mu(b_0),$$

where $b_0' \in B'$. Thus $\pi(b', x) = \pi(b_0', 0) = \psi(b_0')$ and hence $\pi(b', x) \in \operatorname{Im} \psi$. This proves that the sequence

$$0 \to B' \overset{\psi}{\to} X' \overset{\phi}{\to} A \to 0$$

is exact.

We next define a Λ-homomorphism $\tau: X \to X'$ by $\tau(x) = \pi(0, x)$. If now $b \in B$, then

$$\tau g(b) = \pi(0, g(b)) = \pi(\gamma(b), 0) = \psi\gamma(b)$$

and therefore $\tau g = \psi\gamma$. Finally if $x \in X$, then

$$\phi\tau(x) = \phi\pi(0, x) = f(x)$$

which shows that $\phi\tau = f$. This completes the proof.

Exercise 9. *Let* $0 \to B \overset{g}{\to} X \overset{f}{\to} A \to 0$ *be an exact sequence (in* $\mathscr{C}_\Lambda$*) and let* $\omega: A' \to A$ *be a* Λ*-homomorphism. Show that it is possible to construct (in* $\mathscr{C}_\Lambda$*) a commutative diagram*

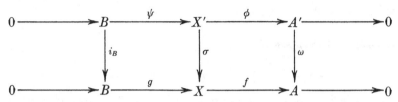

with exact rows.

Theorem 12. *Let* A *and* B *belong to* $\mathscr{C}_\Lambda$*. Then the following statements are equivalent*:

(a) $\mathrm{Ext}^1_\Lambda\,(A, B) = 0$;

(b) *every exact sequence* $0 \to B \to X \to A \to 0$*, in* $\mathscr{C}_\Lambda$*, splits.*

Proof. *Assume* (a) and let $0 \to B \to X \to A \to 0$ be exact. Since
$$\mathrm{Ext}^1_\Lambda\,(A, B) = 0,$$
we have an exact sequence
$$\mathrm{Hom}_\Lambda\,(X, B) \to \mathrm{Hom}_\Lambda\,(B, B) \to 0$$
and therefore i_B is the image of a homomorphism $X \to B$. Accordingly the product of $B \to X$ and $X \to B$ is an identity map and therefore $0 \to B \to X \to A \to 0$ splits.

Assume (b). We first construct an exact sequence $0 \to A_1 \to P \to A \to 0$ with P projective. Let $\gamma \in \mathrm{Hom}_\Lambda\,(A_1, B)$. By Theorem 11, we can produce, in $\mathscr{C}_\Lambda$, a commutative diagram

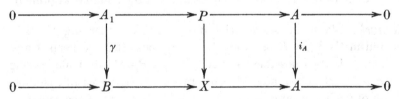

with exact rows, and by hypothesis the lower row splits. Since $\mathrm{Ext}^1_\Lambda\,(P, B) = 0$, Theorem 5 shows that we have a commutative diagram

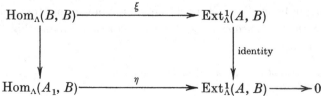

where the lower row is exact and ξ and η are connecting homomorphisms. By Theorem 9, ξ is a null map. Since i_B is in $\text{Hom}_\Lambda(B, B)$ and has image γ in $\text{Hom}_\Lambda(A_1, B)$, it follows that $\eta(\gamma) = 0$. But γ is an arbitrary element of $\text{Hom}_\Lambda(A_1, B)$ and η is surjective. Accordingly $\text{Ext}_\Lambda^1(A, B) = 0$.

Theorem 13. *Suppose that*

$$0 \to A_1 \xrightarrow{\alpha} P \xrightarrow{\beta} A \to 0 \quad and \quad 0 \to B \xrightarrow{\lambda} E \xrightarrow{\mu} B_1 \to 0$$

are exact in $\mathscr{C}_\Lambda$ with P projective and E injective. Then we have an isomorphism $\text{Ext}_\Lambda^1(A_1, B) \approx \text{Ext}_\Lambda^1(A, B_1)$ of abelian groups.

Proof. By Theorems 1 and 2, $\text{Ext}_\Lambda^1(P, B_1) = 0$ and $\text{Ext}_\Lambda^1(A_1, E) = 0$. It follows that we have exact sequences

$$\text{Hom}_\Lambda(A_1, E) \xrightarrow{u} \text{Hom}_\Lambda(A_1, B_1) \to \text{Ext}_\Lambda^1(A_1, B) \to 0$$

and $$\text{Hom}_\Lambda(P, B_1) \xrightarrow{v} \text{Hom}_\Lambda(A_1, B_1) \to \text{Ext}_\Lambda^1(A, B_1) \to 0.$$

The theorem will now follow if we show that $\text{Im}\, u = \text{Im}\, v$.

Let $f \in \text{Im}\, u$. Then there exists a Λ-homomorphism $g : A_1 \to E$ such that $\mu g = f$. Since E is injective, there exists a Λ-homomorphism $\phi : P \to E$ such that $\phi\alpha = g$. Thus $\mu\phi\alpha = f$ and $\mu\phi \in \text{Hom}_\Lambda(P, B_1)$. The relation $\mu\phi\alpha = f$ shows that $f = v(\mu\phi) \in \text{Im}\, v$. This proves that

$$\text{Im}\, u \subseteq \text{Im}\, v$$

and the opposite inclusion can be established by a similar argument.

Remark. We make some brief observations concerning the isomorphism $\text{Ext}_\Lambda^1(A_1, B) \approx \text{Ext}_\Lambda^1(A, B_1)$. Suppose that we keep A and B fixed. It is then possible to vary $0 \to A_1 \to P \to A \to 0$ while leaving $0 \to B \to E \to B_1 \to 0$ unaltered. This reveals that, to within isomorphism of abelian groups, $\text{Ext}_\Lambda^1(A_1, B)$ is independent of the sequence $0 \to A_1 \to P \to A \to 0$. In a similar sense, $\text{Ext}_\Lambda^1(A, B_1)$ is independent of the sequence $0 \to B \to E \to B_1 \to 0$. Let us identify $\text{Ext}_\Lambda^1(A_1, B)$ and $\text{Ext}_\Lambda^1(A, B_1)$ thereby obtaining an abelian group $\text{Ext}_\Lambda^2(A, B)$ say, depending only on A and B. This is, in fact, a new bifunctor called the *second extension functor*. However we shall not pursue this line of investigation.†

---

† For accounts of the higher extension functors see (5), (14) and (19) in the list of references.

Theorem 14. *Let* $0 \to A_n \to P_{n-1} \to \ldots \to P_1 \to P_0 \to A \to 0$ *and*

$$0 \to B \to E_0 \to E_1 \to \ldots \to E_{n-1} \to B_n \to 0$$

be exact sequences in $\mathscr{C}_\Lambda$, *where* $n \geqslant 1$, P_i *is projective and* E_j *injective. Then there is an isomorphism* $\text{Ext}^1_\Lambda (A_n, B) \approx \text{Ext}^1_\Lambda (A, B_n)$ *of abelian groups.*

Proof. Put $A_0 = A$ and, for $1 \leqslant i \leqslant n - 1, A_i = \text{Im} (P_i \to P_{i-1})$. Then we have exact sequences $0 \to A_i \to P_{i-1} \to A_{i-1} \to 0$ for $1 \leqslant i \leqslant n$. Next put $B_0 = B$ and, for $1 \leqslant j \leqslant n - 1$, set $\text{Im} (E_{j-1} \to E_j) = B_j$. This time we have exact sequences $0 \to B_{j-1} \to E_{j-1} \to B_j \to 0$ for $1 \leqslant j \leqslant n$. By Theorem 13 we have an isomorphism

$$\text{Ext}^1_\Lambda (A_{i-1}, B_{n-i+1}) \approx \text{Ext}^1_\Lambda (A_i, B_{n-i})$$

for each i in the range $1 \leqslant i \leqslant n$. The theorem follows.

Theorem 15. *Let* $0 \to A_n \to P_{n-1} \to \ldots \to P_1 \to P_0 \to A \to 0$ *and*

$$0 \to A'_n \to P'_{n-1} \to \ldots \to P'_1 \to P'_0 \to A \to 0$$

be exact sequences in $\mathscr{C}_\Lambda$, *where* $n \geqslant 1$ *and* P_i *and* P'_i *are projective. Then for any* B *in* $\mathscr{C}_\Lambda$ *we have an isomorphism* $\text{Ext}^1_\Lambda (A_n, B) \approx \text{Ext}^1_\Lambda (A'_n, B)$ *of abelian groups. Furthermore* A_n *is projective if and only if* A'_n *is projective.*

Proof. Using (Chapter 2, Theorem 14), we can construct an exact sequence
$$0 \to B \to E_0 \to E_1 \to \ldots \to E_{n-1} \to B_n \to 0,$$

where E_j is injective. Then, by Theorem 14, we have isomorphisms

$$\text{Ext}^1_\Lambda (A_n, B) \approx \text{Ext}^1_\Lambda (A, B_n) \approx \text{Ext}^1_\Lambda (A'_n, B).$$

Finally A_n resp. A'_n is projective if and only if $\text{Ext}^1_\Lambda (A_n, B)$ resp. $\text{Ext}^1_\Lambda (A'_n, B)$ is a null group for all B. The theorem follows.

A very similar argument (which we leave to the reader) will establish

Theorem 16. *Let* $0 \to B \to E_0 \to E_1 \to \ldots \to E_{n-1} \to B_n \to 0$ *and*

$$0 \to B \to E'_0 \to E'_1 \to \ldots \to E'_{n-1} \to B'_n \to 0$$

be exact in $\mathscr{C}_\Lambda$ *with* $n \geqslant 1$ *and* E_i *and* E'_i *injective. Then for any module* A *in* $\mathscr{C}_\Lambda$ *we have an isomorphism* $\text{Ext}^1_\Lambda (A, B_n) \approx \text{Ext}^1_\Lambda (A, B'_n)$ *of abelian groups. Furthermore* B_n *is injective if and only if* B'_n *is injective.*

3.7 Projective and injective dimension

In section (3.7) we shall work primarily in the category $\mathscr{C}_\Lambda^L$ of left Λ-modules.

Suppose that $A \neq 0$ is a left Λ-module. Assume that $n \geqslant 0$ is an integer and that there exists an exact sequence

$$0 \to P_n \to P_{n-1} \to \dots \to P_1 \to P_0 \to A \to 0 \qquad (3.7.1)$$

with each P_i projective, but that there is no exact sequence of the same kind with fewer terms. We then say that A has (left) *projective dimension* n and write $l.\mathrm{Pd}_\Lambda(A) = n$. If no such sequence as (3.7.1) exists we put $l.\mathrm{Pd}_\Lambda(A) = \infty$. Finally, if $A = 0$ then we set

$$l.\mathrm{Pd}_\Lambda(A) = -1.$$

Thus $l.\mathrm{Pd}_\Lambda(A)$ is defined for all modules A in $\mathscr{C}_\Lambda^L$. (If B belongs to $\mathscr{C}_\Lambda^R$ the analogous concept is denoted by $r.\mathrm{Pd}_\Lambda(B)$ and is called the (right) projective dimension of B.) Note that A is projective if and only if $l.\mathrm{Pd}_\Lambda(A) \leqslant 0$. Also isomorphic modules have the same projective dimension. Again if $0 \to A_1 \to P \to A \to 0$ is exact and P is projective, then $l.\mathrm{Pd}_\Lambda(A) = l.\mathrm{Pd}_\Lambda(A_1) + 1$ provided $l.\mathrm{Pd}_\Lambda(A) > 0$.

Exercise 10. *Let $\{A_i\}_{i \in I}$ be a non-empty family of left Λ-modules and A their direct sum. Show that*

$$l.\mathrm{Pd}_\Lambda(A) = \sup_{i \in I} l.\mathrm{Pd}_\Lambda(A_i).$$

Exercise 11.* *Let $0 \to A' \to A \to A'' \to 0$ be an exact sequence of left Λ-modules. Establish the following:*
  (1) *if $l.\mathrm{Pd}_\Lambda(A) > l.\mathrm{Pd}_\Lambda(A')$, then $l.\mathrm{Pd}_\Lambda(A'') = l.\mathrm{Pd}_\Lambda(A)$;*
  (2) *if $l.\mathrm{Pd}_\Lambda(A') > l.\mathrm{Pd}_\Lambda(A)$, then $l.\mathrm{Pd}_\Lambda(A'') = l.\mathrm{Pd}_\Lambda(A') + 1$;*
  (3) *if $l.\mathrm{Pd}_\Lambda(A') = l.\mathrm{Pd}_\Lambda(A)$, then $l.\mathrm{Pd}_\Lambda(A'') \leqslant l.\mathrm{Pd}_\Lambda(A) + 1$.*

If $\phi : \Lambda \to \Delta$ is a ring-homomorphism and U is a left Δ-module, then we can turn U into a left Λ-module by putting $\lambda u = \phi(\lambda)u$ for each λ in Λ and u in U. We recall that a ring is called *non-trivial* if its identity element and zero element are distinct.

Exercise 12. *In the above situation show that, if Δ is a non-trivial ring, then*
$$l.\mathrm{Pd}_\Lambda(U) \leqslant l.\mathrm{Pd}_\Lambda(\Delta) + l.\mathrm{Pd}_\Delta(U).$$

Now suppose that C belongs to $\mathscr{C}_\Lambda^L$ and that $C \neq 0$. Assume that $n \geqslant 0$ is an integer and that there exists an exact sequence

$$0 \to C \to E_0 \to E_1 \to \dots \to E_n \to 0 \qquad (3.7.2)$$

with each E_i injective, but that there is no shorter exact sequence of the same kind. We then say that C has (left) *injective dimension* n and write $l.\mathrm{Id}_\Lambda(C) = n$. If no such sequence as (3.7.2) exists, then we write $l.\mathrm{Id}_\Lambda(C) = \infty$. Finally we put $l.\mathrm{Id}_\Lambda(0) = -1$. Thus $l.\mathrm{Id}_\Lambda(C)$ is defined for all modules C in $\mathscr{C}_\Lambda^L$. (If B belongs to $\mathscr{C}_\Lambda^R$, then the analogous concept is denoted by $r.\mathrm{Id}_\Lambda(B)$ and is called the (right) injective dimension of B.) Note that C is injective if and only if $l.\mathrm{Id}_\Lambda(C) \leqslant 0$. Also isomorphic modules have the same injective dimension. Further if $0 \to C \to E \to C_1 \to 0$ is exact and E is injective, then $l.\mathrm{Id}_\Lambda(C) = l.\mathrm{Id}_\Lambda(C_1) + 1$ provided that $l.\mathrm{Id}_\Lambda(C) > 0$.

Exercise 13. *Determine* $\mathrm{Pd}_Z(Q)$ *and* $\mathrm{Id}_Z(Z)$, *where* Q *denotes the field of rational numbers.*

Exercise 14. *Let* $\{B_i\}_{i \in I}$ *be a non-empty family of left Λ-modules and let B be their direct product. Show that* $l.\mathrm{Id}_\Lambda(B) = \sup_{i \in I} l.\mathrm{Id}_\Lambda(B_i)$.

Theorem 17. *We have*

$$\sup_A l.\mathrm{Pd}_\Lambda(A) = \sup_C l.\mathrm{Id}_\Lambda(C), \qquad (3.7.3)$$

where A and C vary in $\mathscr{C}_\Lambda^L$.

Proof. We first show that

$$\sup_A l.\mathrm{Pd}_\Lambda(A) \leqslant \sup_C l.\mathrm{Id}_\Lambda(C) \qquad (3.7.4)$$

and for this we may suppose that $\sup l.\mathrm{Id}_\Lambda(C) = n$, where $0 \leqslant n < \infty$. Indeed in view of (Chapter 2, Exercise 14) we may assume that $1 \leqslant n < \infty$.

Let A belong to $\mathscr{C}_\Lambda^L$ and, using (Chapter 2, Theorem 6), construct an exact sequence $0 \to A_n \to P_{n-1} \to \dots \to P_1 \to P_0 \to A \to 0$ with P_i projective. If we can show that A_n is projective, this will show $l.\mathrm{Pd}_\Lambda(A) \leqslant n$ and (3.7.4) will follow. Suppose that C belongs to $\mathscr{C}_\Lambda^L$. Observe that if we can show that $\mathrm{Ext}_\Lambda^1(A_n, C) = 0$, then the projective character of A_n will follow from Theorem 1.

To this end we form an exact sequence

$$0 \to C \to E_0 \to E_1 \to \dots \to E_n \to 0,$$

where E_j is injective. This is possible because $l.\mathrm{Id}_\Lambda(C) \leqslant n$. By Theorem 2, $\mathrm{Ext}_\Lambda^1(A, E_n) = 0$. Hence, by Theorem 14, $\mathrm{Ext}_\Lambda^1(A_n, C) = 0$. Thus (3.7.4) is established and the opposite inequality is proved similarly.

In view of Theorem 17 we may put

$$l.\,\mathrm{GD}\,(\Lambda) = \sup_A l.\,\mathrm{Pd}_\Lambda\,(A) = \sup_C l.\,\mathrm{Id}_\Lambda\,(C) \qquad (3.7.5)$$

and call $l.\,\mathrm{GD}\,(\Lambda)$ the (left) *global dimension* of Λ. Here A and C vary in $\mathscr{C}_\Lambda^L$. If we work in $\mathscr{C}_\Lambda^R$ we obtain the (right) global dimension of Λ. This is denoted by $r.\,\mathrm{GD}\,(\Lambda)$. It can happen that $l.\,\mathrm{GD}\,(\Lambda)$ and $r.\,\mathrm{GD}\,(\Lambda)$ are different.†

Theorem 18. *We have*

$$l.\,\mathrm{GD}\,(\Lambda) = \sup_I l.\,\mathrm{Pd}_\Lambda\,(\Lambda/I), \qquad (3.7.6)$$

where I ranges over all the left ideals of Λ.

Proof. It is enough to show that

$$l.\,\mathrm{GD}\,(\Lambda) \leqslant \sup_I l.\,\mathrm{Pd}_\Lambda\,(\Lambda/I) \qquad (3.7.7)$$

since the opposite inequality is obvious. For this we may suppose that $\sup_I l.\,\mathrm{Pd}_\Lambda\,(\Lambda/I) = n$, where $0 \leqslant n < \infty$.

Let C belong to $\mathscr{C}_\Lambda^L$. It is sufficient to prove that $l.\,\mathrm{Id}_\Lambda\,(C) \leqslant n$. If $n = 0$, then, for every I, Λ/I is projective and hence $\mathrm{Ext}_\Lambda^1\,(\Lambda/I, C) = 0$. It follows from Theorem 8, that C is injective and therefore

$$l.\,\mathrm{Id}_\Lambda\,(C) \leqslant 0 \leqslant n.$$

We may therefore suppose that $1 \leqslant n < \infty$. Construct, with the aid of (Chapter 2, Theorem 14), an exact sequence

$$0 \to C \to E_0 \to \ldots \to E_{n-1} \to C_n \to 0$$

with E_i injective. It now suffices to show that C_n is injective. By Theorem 8, this will follow if we show that $\mathrm{Ext}_\Lambda^1\,(\Lambda/I, C_n) = 0$ for each left ideal I.

Let I be a left ideal. Since $l.\,\mathrm{Pd}_\Lambda\,(\Lambda/I) \leqslant n$ we can construct an exact sequence
$$0 \to P_n \to P_{n-1} \to \ldots \to P_0 \to \Lambda/I \to 0$$

with P_i projective for $0 \leqslant i \leqslant n$. Then $\mathrm{Ext}_\Lambda^1\,(P_n, C) = 0$. But now $\mathrm{Ext}_\Lambda^1\,(\Lambda/I, C_n) = 0$ by Theorem 14, and with this the proof is complete.

Exercise 15. *Show that* $\mathrm{GD}\,(Z) = 1$.

Let A be a left Λ-module. We know that we can construct an exact sequence $0 \to A_1 \to P_0 \to A \to 0$ with P_0 projective. We may then construct an exact sequence $0 \to A_2 \to P_1 \to A_1 \to 0$ with P_1 projective.

† See I. Kaplansky (12).

Indeed it is possible to continue this process indefinitely, the typical exact sequence having the form $0 \to A_{i+1} \to P_i \to A_i \to 0$, P_i being projective. By combining these short exact sequences we arrive at an *infinite* exact sequence

$$\ldots \to P_m \to P_{m-1} \to \ldots \to P_1 \to P_0 \to A \to 0$$

where P_0, P_1, P_2 etc. are all projective. An exact sequence of this form is called a *projective resolution* of A. Note that A will have finite projective dimension if and only if it has a projective resolution which is composed of zero modules from some point onwards.

Now let B be a left Λ-module. Using similar considerations we can construct an *injective resolution* of B, that is to say an infinite exact sequence

$$0 \to B \to E_0 \to E_1 \to \ldots \to E_{m-1} \to E_m \to \ldots$$

where each E_i is injective. Of course B has finite injective dimension if and only if it has an injective resolution whose component modules are zero modules from some point onwards.

However, we can do rather better in the injective case. Let E_0 be the injective envelope† of B. Then we can construct an exact sequence $0 \to B \to E_0 \to B_1 \to 0$, where $B \to E_0$ is an inclusion mapping. In the same way we can construct an exact sequence $0 \to B_1 \to E_1 \to B_2 \to 0$, where E_1 is the injective envelope of B_1, and so on indefinitely. These sequences can then be put together to give an injective resolution of B. In particular we see that there exists an exact sequence

$$0 \to B \overset{\epsilon}{\to} E_0 \overset{d_0}{\to} E_1 \overset{d_1}{\to} E_2 \overset{d_2}{\to} E_3 \to \ldots \qquad (3.7.8)$$

where (i) E_0 is the injective envelope of $\epsilon(B)$, and (ii) for each $i \geqslant 0$, E_{i+1} is the injective envelope of $d_i(E_i)$. Such an exact sequence is known as a *minimal injective resolution* of B.

Theorem 19. *Let the Λ-modules B and B' have minimal injective resolutions*

$$0 \to B \to E_0 \overset{d_0}{\to} E_1 \overset{d_1}{\to} E_2 \to \ldots$$

and

$$0 \to B' \to E_0' \overset{d_0'}{\to} E_1' \overset{d_1'}{\to} E_2' \to \ldots$$

respectively. Further let $f \colon B \overset{\sim}{\to} B'$ be an isomorphism in $\mathscr{C}_\Lambda$. Then it is possible to construct, in succession, Λ-isomorphisms $\phi_0 \colon E_0 \overset{\sim}{\to} E_0'$,

† See section (2.6).

$\phi_1\colon E_1 \xrightarrow{\sim} E_1'$ *and so on which are such that the diagram*

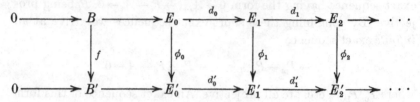

is commutative.

Remark. This theorem shows that minimal injective resolutions are essentially unique.

Proof. The existence of ϕ_0 making the first square commutative follows from (Chapter 2, Theorem 18). This situation induces an isomorphism of $d_0(E_0)$ on to $d_0'(E_0')$. By the same theorem, the induced isomorphism of $d_0(E_0)$ on to $d_0'(E_0')$ can be extended to an isomorphism $\phi_1\colon E_1 \xrightarrow{\sim} E_1'$ between their injective envelopes, and now we have $\phi_1 d_0 = d_0' \phi_0$. Next there is induced an isomorphism of $d_1(E_1)$ on to $d_1'(E_1')$ and this leads to an isomorphism $\phi_2\colon E_2 \xrightarrow{\sim} E_2'$ satisfying

$$\phi_2 d_1 = d_1' \phi_1.$$

And so on indefinitely.

Theorem 20. *Suppose that*

$$0 \to B \to E_0 \xrightarrow{d_0} E_1 \xrightarrow{d_1} E_2 \to \dots$$

is a minimal injective resolution of the left Λ-module B and let $n \geqslant 0$ be an integer. Then $l.\mathrm{Id}_\Lambda(B) < n$ if and only if $E_i = 0$ for all $i \geqslant n$.

Proof. It is clear that if $E_m = 0$, then $E_h = 0$ for all $h \geqslant m$. We shall therefore assume that $l.\mathrm{Id}_\Lambda(B) < n$ and deduce that $E_n = 0$. The proof will then be complete since the other assertions are trivial. It is evident that we may suppose that $n \geqslant 2$ for this part of the argument.

We have an exact sequence

$$0 \to B \to E_0 \to E_1 \to \dots \to E_{n-2} \to d_{n-2}(E_{n-2}) \to 0$$

and, since $l.\mathrm{Id}_\Lambda(B) < n$, there also exists an exact sequence

$$0 \to B \to E_0' \to E_1' \to \dots \to E_{n-2}' \to E_{n-1}' \to 0,$$

where each E_j' is injective. By Theorem 16, $d_{n-2}(E_{n-2})$ must also be injective and this implies that $d_{n-2}(E_{n-2}) = E_{n-1}$. Thus d_{n-2} is surjective and therefore $d_{n-1}(E_{n-1}) = 0$. Since E_n is the injective envelope of $d_{n-1}(E_{n-1})$, this proves that $E_n = 0$.

Exercise 16.* *Suppose that*

$$0 \to B \to E_0 \to E_1 \to E_2 \to \dots$$

is a minimal injective resolution of the left Λ-module B and let γ be an element of the centre of Λ which is not a zerodivisor† on either Λ or B. Show that if $D = \mathrm{Im}\,(E_0 \to E_1)$, then

$$0 \to \mathrm{Hom}_\Lambda\,(\Lambda/\gamma\Lambda, D) \to \mathrm{Hom}_\Lambda\,(\Lambda/\gamma\Lambda, E_1) \to \mathrm{Hom}_\Lambda\,(\Lambda/\gamma\Lambda, E_2) \to \dots$$

is a minimal injective resolution of the $\Lambda/\gamma\Lambda$-module $\mathrm{Hom}_\Lambda\,(\Lambda/\gamma\Lambda, D)$. Show also that $\mathrm{Hom}_\Lambda\,(\Lambda/\gamma\Lambda, D)$ and $B/\gamma B$ are isomorphic $\Lambda/\gamma\Lambda$-modules.

Exercise 17. *Let B be a left Λ-module and γ an element of the centre of Λ which is not a zerodivisor on either Λ or B. Suppose that $l.\mathrm{Id}_\Lambda\,(B) < n$, where $1 \leqslant n < \infty$. Show that $l.\mathrm{Id}_{\Lambda/\gamma\Lambda}\,(B/\gamma B) < n - 1$.*

3.8 Λ-sequences

We begin with an exercise.

Exercise. 18 *Let I be a two-sided ideal of Λ and P a projective left Λ-module. Show that P/IP is a projective left Λ/I-module.*

Theorem 21. *Let γ be an element in the centre of Λ which is neither a unit nor a zerodivisor, and let A be a left $\Lambda/\gamma\Lambda$-module such that*

$$l.\mathrm{Pd}_{\Lambda/\gamma\Lambda}\,(A) = n,$$

where $0 \leqslant n < \infty$. Then $l.\mathrm{Pd}_\Lambda\,(A) = n + 1$.

Remark. It is not the case that if K is a left $\Lambda/\gamma\Lambda$-module such that $l.\mathrm{Pd}_{\Lambda/\gamma\Lambda}\,(K) = \infty$, then necessarily $l.\mathrm{Pd}_\Lambda\,(K) = \infty$ as well.

Proof. We use induction on n. First suppose that $n = 0$. Then A is a direct summand of a non-zero free $\Lambda/\gamma\Lambda$-module Φ and it remains a direct summand of Φ if we regard both A and Φ as Λ-modules. Hence (Exercise 10) $l.\mathrm{Pd}_\Lambda\,(A) \leqslant l.\mathrm{Pd}_\Lambda\,(\Phi)$. But, as a Λ-module, Φ is a direct sum of copies of $\Lambda/\gamma\Lambda$. Accordingly, again by Exercise 10, $l.\mathrm{Pd}_\Lambda\,(\Phi) = l.\mathrm{Pd}_\Lambda\,(\Lambda/\gamma\Lambda)$. But, because γ is not a zerodivisor, we have an exact sequence $0 \to \Lambda \xrightarrow{\gamma} \Lambda \to \Lambda/\gamma\Lambda \to 0$ of Λ-modules and therefore, by Exercise 11,

$$l.\mathrm{Pd}_\Lambda\,(\Lambda/\gamma\Lambda) \leqslant l.\mathrm{Pd}_\Lambda\,(\Lambda) + 1 = 1.$$

---

† The assertion that γ is not a zerodivisor on B means that if $b \in B$ and $\gamma b = 0$, then b must be zero.

Thus for any free $\Lambda/\gamma\Lambda$-module Φ we have

$$l.\operatorname{Pd}_\Lambda(\Phi) \leqslant 1$$

and, in the present instance, we also have $0 \leqslant l.\operatorname{Pd}_\Lambda(A) \leqslant 1$. Now we cannot have $l.\operatorname{Pd}_\Lambda(A) = 0$. (For in that case A would be a submodule of a free Λ-module F and γ would not annihilate any non-zero element of F. However $A \neq 0$ and $\gamma A = 0$. Thus we arrive at a contradiction.) Accordingly $l.\operatorname{Pd}_\Lambda(A) = 1$ and the theorem has been proved for the case $n = 0$.

We now turn to the case $n = 1$. In this situation we can construct an exact sequence $0 \to L \to M \to A \to 0$ of $\Lambda/\gamma\Lambda$-modules, where L and M are $\Lambda/\gamma\Lambda$-projective. Accordingly M is a direct summand of a free $\Lambda/\gamma\Lambda$-module Φ say, and therefore $l.\operatorname{Pd}_\Lambda(M) \leqslant l.\operatorname{Pd}_\Lambda(\Phi) \leqslant 1$ as we saw above. Similarly $l.\operatorname{Pd}_\Lambda(L) \leqslant 1$. Consequently

$$0 \leqslant l.\operatorname{Pd}_\Lambda(A) \leqslant 2$$

by virtue of Exercise 11. However we cannot have $l.\operatorname{Pd}_\Lambda(A) = 0$ for the same reason that this was ruled out in the case $n = 0$. It will now be assumed that $l.\operatorname{Pd}_\Lambda(A) = 1$ and from this we shall derive a contradiction. The theorem will then be proved when $n = 1$.

We can construct an exact sequence $0 \to P \to F \to A \to 0$ of Λ-modules, where F is free and P is a submodule of F. Since we are supposing that $l.\operatorname{Pd}_\Lambda(A) = 1$, P is projective. Now $\gamma F \subseteq P$ because $\gamma A = 0$ and therefore we have an exact sequence

$$0 \to \frac{P}{\gamma F} \to \frac{F}{\gamma F} \to A \to 0$$

in $\mathscr{C}_{\Lambda/\gamma\Lambda}$. By Exercise 18, $F/\gamma F$ is $\Lambda/\gamma\Lambda$-projective and therefore $P/\gamma F$ is $\Lambda/\gamma\Lambda$-projective because $n = 1$. It follows that the exact sequence

$$0 \to \frac{\gamma F}{\gamma P} \to \frac{P}{\gamma P} \to \frac{P}{\gamma F} \to 0,$$

in $\mathscr{C}_{\Lambda/\gamma\Lambda}$, splits. However, by Exercise 18, $P/\gamma P$ is $\Lambda/\gamma\Lambda$-projective. Consequently $\gamma F/\gamma P$ is $\Lambda/\gamma\Lambda$-projective as well. But multiplication by γ induces an isomorphism $F \approx \gamma F$ in which P corresponds to γP. Thus there result $\Lambda/\gamma\Lambda$-isomorphisms

$$A \approx F/P \approx \gamma F/\gamma P$$

which imply that A is $\Lambda/\gamma\Lambda$-projective, contrary to our assumption that $n = 1$.

The theorem has now been established both when $n = 0$ and $n = 1$. We therefore suppose that $n \geqslant 2$ and assume the result to be known for all smaller values of the inductive variable. In $\mathscr{C}_{\Lambda/\gamma\Lambda}$ we construct an exact sequence $\quad 0 \to A_1 \to \Phi \to A \to 0,$

where Φ is $\Lambda/\gamma\Lambda$-free. Then $l.\mathrm{Pd}_{\Lambda/\gamma\Lambda}(A_1) = n - 1$ and therefore, by the inductive hypothesis, $l.\mathrm{Pd}_\Lambda(A_1) = n$. On the other hand,

$$l.\mathrm{Pd}_\Lambda(\Phi) \leqslant 1$$

by our previous observations. It therefore follows, by Exercise 11, that $l.\mathrm{Pd}_\Lambda(A) = l.\mathrm{Pd}_\Lambda(A_1) + 1 = n + 1$ and with this the proof is complete.

Definition. *A sequence* $\gamma_1, \gamma_2, \ldots, \gamma_n$ *of elements belonging to the centre of* Λ *will be called a 'Λ-sequence' if for each i $(1 \leqslant i \leqslant n)$ γ_i is not a zero-divisor on* $\Lambda/(\gamma_1\Lambda + \gamma_2\Lambda + \ldots + \gamma_{i-1}\Lambda)$.

Suppose that $\gamma_1, \gamma_2, \ldots, \gamma_n$ $(n \geqslant 2)$ is a Λ-sequence. Then γ_1 is not a zerodivisor. Further, if Λ^* is the ring $\Lambda/\gamma_1\Lambda$ and γ_i^* is the natural image of γ_i in Λ^*, then the elements $\gamma_2^*, \gamma_3^*, \ldots, \gamma_n^*$ belong to the centre of Λ^* and they form a Λ^*-sequence.

Theorem 22. *Let* $\gamma_1, \gamma_2, \ldots, \gamma_n$ $(n \geqslant 1)$ *belong to the centre of* Λ *and form a Λ-sequence. If now* $\gamma_1\Lambda + \gamma_2\Lambda + \ldots + \gamma_n\Lambda \neq \Lambda$, *then*

$$l.\mathrm{Pd}_\Lambda\{\Lambda/(\gamma_1\Lambda + \gamma_2\Lambda + \ldots + \gamma_n\Lambda)\} = n.$$

Proof. We use induction on n. First, however, we note that γ_1 is neither a zerodivisor nor a unit.

If $n = 1$, then because $l.\mathrm{Pd}_{\Lambda/\gamma_1\Lambda}(\Lambda/\gamma_1\Lambda) = 0$ Theorem 21 shows that $l.\mathrm{Pd}_\Lambda(\Lambda/\gamma_1\Lambda) = 1$ as required. We shall therefore suppose that $n \geqslant 2$ and that the theorem has been proved for Λ-sequences with only $n - 1$ terms.

It follows, from the induction hypothesis, that if $\Lambda^* = \Lambda/\gamma_1\Lambda$, then $l.\mathrm{Pd}_{\Lambda^*}\{\Lambda^*/(\gamma_2^*\Lambda^* + \ldots + \gamma_n^*\Lambda^*)\} = n - 1$. Accordingly, by Theorem 21,

$$l.\mathrm{Pd}_\Lambda\{\Lambda^*/(\gamma_2^*\Lambda^* + \ldots + \gamma_n^*\Lambda^*)\} = n.$$

However $\Lambda^*/(\gamma_2^*\Lambda^* + \ldots + \gamma_n^*\Lambda^*)$ and $\Lambda/(\gamma_1\Lambda + \gamma_2\Lambda + \ldots + \gamma_n\Lambda)$ are isomorphic Λ-modules and so the desired result follows.

3.9 The extension problem

Let A and B belong to $\mathscr{C}_\Lambda$.

Definition. *By an 'extension of B by A' we mean an exact sequence*

$$(S) \quad 0 \to B \to X \to A \to 0 \qquad (3.9.1)$$

in $\mathscr{C}_\Lambda$.

The extension problem is to classify all such extensions. Suppose therefore that we have a second extension of B by A say

$$(S') \quad 0 \to B \to X' \to A \to 0. \tag{3.9.2}$$

We shall say that S and S' are *equivalent* extensions if there exists a Λ-isomorphism $f: X \overset{\sim}{\to} X'$ which makes the diagram

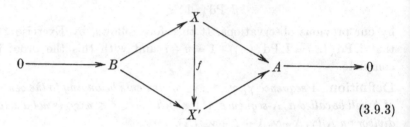

$$\tag{3.9.3}$$

commutative. It is clear that this gives us a relation between extensions which is reflexive, symmetric and transitive. We therefore speak of classes of equivalent extensions.

Exercise 19. *Suppose that* $0 \to B \to X \to A \to 0$ *and* $0 \to B \to X' \to A \to 0$ *are exact, and that* $f: X \to X'$ *is a* Λ-*homomorphism which makes the diagram*

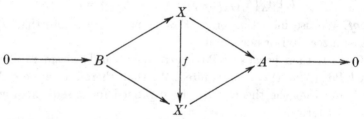

commutative. Show that f *is an isomorphism.*

Consideration of the canonical exact sequence

$$0 \to B \to B \oplus A \to A \to 0$$

shows that there exist extensions of B by A which are split exact sequences. If an extension splits, then so does any extension which is equivalent to it. If two extensions both split, then they are equivalent. Accordingly there is a class of extensions of B by A all of whose members split and which contains all split extensions.

Suppose that we have an extension

$$(S) \quad 0 \to B \overset{\psi}{\to} X \overset{\phi}{\to} A \to 0. \tag{3.9.4}$$

This will give rise to a connecting homomorphism

$$\Delta^1_S \colon \mathrm{Hom}_\Lambda\,(B, B) \to \mathrm{Ext}^1_\Lambda\,(A, B). \qquad (3.9.5)$$

Furthermore, by Theorem 5, if S' is an extension equivalent to S, then the mappings Δ^1_S and $\Delta^1_{S'}$ will coincide. It follows that if we put

$$\Delta^1(S) = \Delta^1_S(i_B), \qquad (3.9.6)$$

then $\Delta^1(S)$ belongs to $\mathrm{Ext}^1_\Lambda\,(A, B)$ and we have, in effect, a mapping of the classes of equivalent extensions of B by A into the abelian group $\mathrm{Ext}^1_\Lambda\,(A, B)$. We shall show, in a moment, that this mapping is a bijection, but first we shall make one or two preliminary remarks.

To begin with we observe that if S is a split exact sequence, then, by Theorem 9, Δ^1_S is a null homomorphism. Accordingly, *the element of* $\mathrm{Ext}^1_\Lambda\,(A, B)$ *which arises from the class of split extensions is the zero element.* Again we could use (3.9.4) to obtain a connecting homomorphism

$$\Delta^2_S \colon \mathrm{Hom}_\Lambda\,(A, A) \to \mathrm{Ext}^1_\Lambda\,(A, B),$$

and just as we put $\Delta^1(S) = \Delta^1_S(i_B)$ we could equally well put

$$\Delta^2(S) = \Delta^2_S(i_A)$$

and this would give us an alternative method of proceeding. However this would yield nothing new because of

Exercise 20. *Show that if S is an extension of B by A, then*

$$\Delta^1(S) = \Delta^2(S).$$

In view of the result contained in this exercise, we shall forget all about $\Delta^2(S)$ and, to simplify the notation, we shall write Δ_S and $\Delta(S)$ in place of Δ^1_S and $\Delta^1_S(i_B)$ respectively.

We have to show that the association of $\Delta(S)$ with S provides us with a bijection as stated above. To this end we construct an exact sequence $0 \to A_1 \xrightarrow{\beta} P \xrightarrow{\alpha} A \to 0$ with P projective. In what follows this sequence will be kept *fixed*. Suppose now that we have an extension (3.9.4). Then, because P is projective, we can construct, in $\mathscr{C}_\Lambda$, a commutative diagram

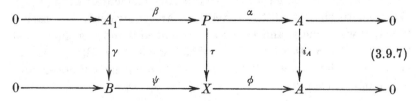

(3.9.7)

which, by Theorem 5, yields a commutative diagram

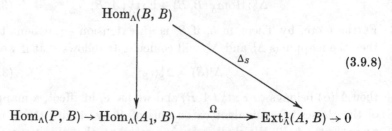

$$(3.9.8)$$

where the row is exact and Ω is the connecting homomorphism associated with the upper row of (3.9.7). Since the image of i_B in

$$\mathrm{Hom}_\Lambda (A_1, B)$$

is γ, it follows that $\quad\quad \Omega(\gamma) = \Delta_S(i_B) = \Delta(S)$. $\quad\quad\quad$ (3.9.9)

Now suppose that we are given an arbitrary Λ-homomorphism $\gamma : A_1 \to B$. By Theorem 11, we can then construct a commutative diagram (3.9.7), and therefore we can find an extension S such that $\Delta(S) = \Omega(\gamma)$. But Ω is surjective. Consequently every element of $\mathrm{Ext}^1_\Lambda (A, B)$ has the form $\Delta(S)$.

Let us return to the diagram (3.9.7). Each element $x \in X$ can be expressed in the form $x = \tau(p) + \psi(b)$, where $p \in P$ and $b \in B$. Suppose that $x = \tau(p') + \psi(b')$ is a second such representation. Then

$$\tau(p - p') = \psi(b' - b),$$

whence $\alpha(p - p') = \phi\tau(p - p') = 0$. Thus there exists $a_1 \in A_1$ such that $p - p' = \beta(a_1)$ and

$$\psi\gamma(a_1) = \tau\beta(a_1) = \tau(p - p') = \psi(b' - b).$$

But ψ is monic. *Consequently* $\tau(p) + \psi(b) = \tau(p') + \psi(b')$ *implies that* $p - p' = \beta(a_1)$ *and* $b - b' = -\gamma(a_1)$ *for some* $a_1 \in A_1$.

Finally assume that we have two extensions, say

$$0 \to B \overset{\psi_1}{\to} X_1 \overset{\phi_1}{\to} A \to 0 \quad \text{and} \quad 0 \to B \overset{\psi_2}{\to} X_2 \overset{\phi_2}{\to} A \to 0,$$

and suppose that these give rise to the same element of $\mathrm{Ext}^1_\Lambda (A, B)$. We wish to show that the extensions are equivalent. For each i ($i = 1, 2$) we define γ_i and τ_i just as we defined γ and τ in the case of the extension S. It then follows from (3.9.9) that $\Omega(\gamma_1) = \Omega(\gamma_2)$. However the lower row of (3.9.8) is exact. Consequently there exists a Λ-homomorphism $\theta : P \to B$ such that $\theta\beta = \gamma_1 - \gamma_2$.

We now define a mapping $f: X_1 \to X_2$. Each element of X_1 can be expressed in the form $\tau_1(p) + \psi_1(b)$. We put

$$f(\tau_1(p) + \psi_1(b)) = \tau_2(p) + \psi_2(b) + \psi_2\theta(p)$$

and then an easy verification (using the results of the last paragraph but one) shows that f is well defined. In fact f is a Λ-homomorphism and, since

$$\phi_1(\tau_1(p) + \psi_1(b)) = \phi_1\tau_1(p) = \alpha(p)$$

and

$$\phi_2(\tau_2(p) + \psi_2(b) + \psi_2\theta(p)) = \phi_2\tau_2(p) = \alpha(p),$$

the diagram

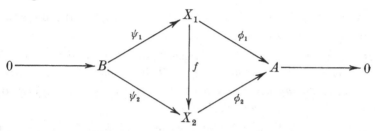

is commutative. It follows (Exercise 19) that f is an isomorphism. Accordingly the two extensions are equivalent and we have proved

Theorem 23. *There is a bijection between the classes of equivalent extensions of B by A and the elements of $\mathrm{Ext}_\Lambda^1(A, B)$ in which the class of the extension S corresponds to $\Delta_S^1(i_B)$. Moreover the class of split extensions corresponds to the zero element of $\mathrm{Ext}_\Lambda^1(A, B)$.*

Exercise 21.* *Let $0 \to B \to X_1 \to A \to 0$ and $0 \to B \to X_2 \to A \to 0$ be two extensions of B by A and suppose that they correspond to the elements ω_1 and ω_2 respectively of $\mathrm{Ext}_\Lambda^1(A, B)$. Construct an extension of B by A that corresponds to $\omega_1 + \omega_2$.*

Solutions to the Exercises on Chapter 3

Exercise 1. *Let B be a Λ-module. Show that the following two statements are equivalent:*

(a) B is injective;

(b) $\mathrm{Ext}_\Lambda^1(A, B) = 0$ for all A in $\mathscr{C}_\Lambda$.

Solution. By Lemma 2, (a) implies (b). Assume (b) and construct an exact sequence $0 \to B \to E \to B_1 \to 0$, where E is injective. Since

$$\mathrm{Ext}_\Lambda^1(B_1, B) = 0,$$

we obtain an exact sequence

$$0 \to \operatorname{Hom}_\Lambda (B_1, B) \to \operatorname{Hom}_\Lambda (B_1, E) \to \operatorname{Hom}_\Lambda (B_1, B_1) \to 0.$$

Hence there exists $f \in \operatorname{Hom}_\Lambda (B_1, E)$ such that the diagram

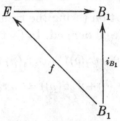

is commutative. Accordingly $0 \to B \to E \to B_1 \to 0$ splits. It follows that B is a direct summand of E and so it is injective.

Exercise 2. *Suppose that A, B belong to $\mathscr{C}_\Lambda$ and that Γ is the centre of Λ. Since A is a (Λ, Γ)-bimodule, $\operatorname{Ext}^1_\Lambda (A, B)$ is a Γ-module. Since B is a (Λ, Γ)-bimodule, this also induces a Γ-module structure on $\operatorname{Ext}^1_\Lambda (A, B)$. Show that these two structures coincide.*

Solution. Let γ belong to Γ and, for each Λ-module C, let $\gamma_C : C \to C$ be the Λ-homomorphism which results from multiplication by γ. We have to show that $\operatorname{Ext}^1_\Lambda (\gamma_A, B) = \operatorname{Ext}^1_\Lambda (A, \gamma_B)$.

Construct an exact sequence $0 \to B \to E \to B_1 \to 0$ with E injective. Then the diagram

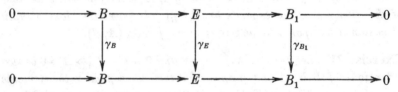

is commutative and, by the definition of $\operatorname{Ext}^1_\Lambda$, we have commutative diagrams

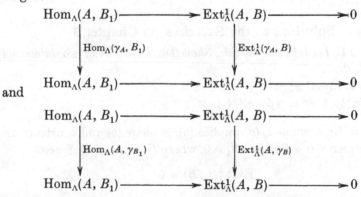

and

where the rows are exact. However, by (Chapter 2, Exercise 2), $\mathrm{Hom}_\Lambda(\gamma_A, B_1) = \mathrm{Hom}_\Lambda(A, \gamma_{B_1})$ and from this the desired result follows.

Exercise 3. *Let*

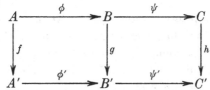

be a commutative diagram in $\mathscr{C}_\Lambda$ with exact rows. Show that if ϕ' is monic, then the sequence $\mathrm{Ker}\, f \to \mathrm{Ker}\, g \to \mathrm{Ker}\, h$ is exact, whereas if ψ is epic, then $\mathrm{Coker}\, f \to \mathrm{Coker}\, g \to \mathrm{Coker}\, h$ is exact.

Solution. The first assertion was proved in the discussion of Lemma 4. Also, in any event, it is clear that the product of $\mathrm{Coker}\, f \to \mathrm{Coker}\, g$ and $\mathrm{Coker}\, g \to \mathrm{Coker}\, h$ is null.

Now assume that ψ is epic. Suppose that $b' \in B'$ and let $[b']$ denote its image in $\mathrm{Coker}\, g$. If $[b']$ maps into zero in $\mathrm{Coker}\, h$, then $\psi'(b') = h(c)$ for some $c \in C$ and hence, because ψ is epic, $\psi'(b') = h\psi(b) = \psi'g(b)$ for a suitable b in B. It follows that $b' - g(b)$ is in $\mathrm{Ker}\, \psi' = \mathrm{Im}\, \phi'$. Consequently there exists $a' \in A'$ such that $b' = g(b) + \phi'(a')$. Accordingly $[b'] = [\phi'(a')]$ that is to say $[b']$ is the image of the element $[a']$, of $\mathrm{Coker}\, f$, that corresponds to a'. This shows that

$$\mathrm{Ker}\,\{\mathrm{Coker}\, g \to \mathrm{Coker}\, h\} \subseteq \mathrm{Im}\,\{\mathrm{Coker}\, f \to \mathrm{Coker}\, g\}$$

and completes the argument.

Exercise 4. *Let*

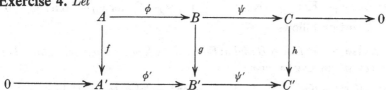

be a commutative diagram in $\mathscr{C}_\Lambda$ with exact rows. Show that the Ker–Coker sequence

$$\mathrm{Ker}\, f \to \mathrm{Ker}\, g \to \mathrm{Ker}\, h \overset{\Delta}{\to} \mathrm{Coker}\, f \to \mathrm{Coker}\, g \to \mathrm{Coker}\, h$$

is exact.

Solution. It remains to be shown that $\mathrm{Ker}\, h \overset{\Delta}{\to} \mathrm{Coker}\, f \to \mathrm{Coker}\, g$ is exact (see Theorem 4, Lemma 4, and Exercise 3). Suppose therefore that $c \in \mathrm{Ker}\, h$. Since ψ is epic, there exists $b \in B$ such that $\psi(b) = c$, and then $\psi'g(b) = h\psi(b) = h(c) = 0$. Hence $g(b)$ is in $\operatorname{Ker}\psi' = \operatorname{Im}\phi'$, that is $g(b) = \phi'(a')$ for some a' in A'. By the definition of the connecting homomorphism $\Delta(c) = [a']$, where the notation is the same as in Exercise 3. But $[\phi'(a')] = [g(b)] = 0$ and therefore c maps into zero under the composite homomorphism $\operatorname{Ker} h \overset{\Delta}{\to} \operatorname{Coker} f \to \operatorname{Coker} g$.

Suppose next that $[a']$ maps into zero under $\operatorname{Coker} f \to \operatorname{Coker} g$. Then $\phi'(a') = g(b)$ for some b in B and $h\psi(b) = \psi'g(b) = \psi'\phi'(a') = 0$. Thus $\psi(b) \in \operatorname{Ker} h$ and $\Delta\psi(b) = [a']$. It follows that $[a']$ is in $\operatorname{Im}\Delta$ and with this the solution is complete.

Exercise 5. *Suppose that the sequence* $0 \to A \to B \to C \to 0$ *is exact in* $\mathscr{C}_\Lambda$ *and let* γ *be a central element of* Λ. *Establish the existence of an exact sequence*

$$0 \to (0:_A \gamma) \to (0:_B \gamma) \to (0:_C \gamma) \to A/\gamma A \to B/\gamma B \to C/\gamma C \to 0$$

where, for example, $0:_A \gamma$ *is the submodule of* A *annihilated by* γ.

Solution. For each M in $\mathscr{C}_\Lambda$ let $\gamma_M : M \to M$ be the Λ-homomorphism which results from multiplication by γ. Then $\operatorname{Ker}\gamma_M = 0:_M \gamma$ and $\operatorname{Coker}\gamma_M = M/\gamma M$. Now

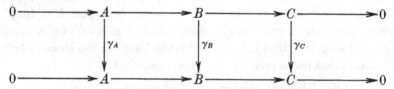

is a commutative diagram with exact rows. Hence, by Theorem 4 and the note following it, we obtain an exact sequence

$$0 \to \operatorname{Ker}\gamma_A \to \operatorname{Ker}\gamma_B \to \operatorname{Ker}\gamma_C \to \operatorname{Coker}\gamma_A \to \operatorname{Coker}\gamma_B \to \operatorname{Coker}\gamma_C \to 0.$$

The assertion follows.

Exercise 6. *Let* $f:A \to B$ *and* $g:B \to C$ *be* Λ-*homomorphisms. Show that there exists an exact sequence*

$$0 \to \operatorname{Ker} f \to \operatorname{Ker}(gf) \to \operatorname{Ker} g \to \operatorname{Coker} f \to \operatorname{Coker}(gf) \to \operatorname{Coker} g \to 0.$$

Solution. Define a Λ-homomorphism $h : A \oplus B \to B \oplus C$ by

$$h(a, b) = (f(a) - b, g(b)).$$

Then

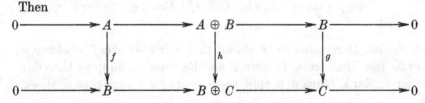

is a commutative diagram with exact rows, where the horizontal mappings are the natural ones. Accordingly, by Theorem 4 and the note following it, we have an exact sequence

$$0 \to \operatorname{Ker} f \to \operatorname{Ker} h \to \operatorname{Ker} g \to \operatorname{Coker} f \to \operatorname{Coker} h \to \operatorname{Coker} g \to 0.$$

We shall complete the proof by showing that $\operatorname{Ker} h$ resp. $\operatorname{Coker} h$ is isomorphic to $\operatorname{Ker} (gf)$ resp. $\operatorname{Coker} (gf)$.

If $h(a, b) = 0$, then $g(b) = 0$ and $f(a) = b$. Thus $gf(a) = 0$ and hence $a \in \operatorname{Ker} (gf)$. We can therefore define a homomorphism

$$\phi : \operatorname{Ker} h \to \operatorname{Ker} (gf)$$

by $\phi(a, b) = a$. Clearly ϕ is monic. Also if $\alpha \in \operatorname{Ker} (gf)$, then

$$(\alpha, f(\alpha)) \in \operatorname{Ker} h$$

and is mapped, by ϕ, into α. This shows that ϕ maps $\operatorname{Ker} h$ isomorphically on to $\operatorname{Ker} (gf)$.

Let $\mu : B \oplus C \to C$ be the Λ-homomorphism defined by

$$\mu(b, c) = g(b) + c.$$

If $(b, c) \in \operatorname{Im} h$, then there exist $\alpha \in A$ and $\beta \in B$ such that

$$(b, c) = (f(\alpha) - \beta, g(\beta))$$

and therefore $\mu(b, c) = gf(\alpha)$. Thus $\mu(\operatorname{Im} h) \subseteq \operatorname{Im} (gf)$ and hence μ induces a homomorphism $\psi : \operatorname{Coker} h \to \operatorname{Coker} (gf)$ which is epic because μ is epic. We shall now show that ψ is also monic and with this the solution will be complete.

Suppose that the image of (b, c) in $\operatorname{Coker} h$ is mapped into zero by ψ, i.e. suppose that $\mu(b, c) = gf(a)$ for some $a \in A$. It will suffice to show that $(b, c) \in \operatorname{Im} h$. But $gf(a) = \mu(b, c) = g(b) + c$ whence

$$c = g(f(a) - b).$$

Thus $h(a, f(a) - b) = (b, c)$ and consequently $(b, c) \in \operatorname{Im} h$ as required.

Exercise 7. *Let Γ be the centre of Λ. Show that the connecting homomorphisms (denoted by Δ) in Theorems 5 and 6 are Γ-homomorphisms.*

Solution. For any Λ-module M and $\gamma \in \Gamma$, denote by $\gamma_M : M \to M$ the homomorphism which multiplies every element of M by γ.

Consider an exact sequence $0 \to A' \to A \to A'' \to 0$. Then

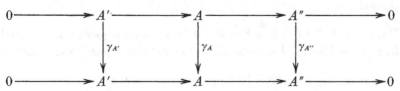

is a commutative diagram in $\mathscr{C}_\Lambda$ with exact rows. From Theorem 5 we obtain a commutative diagram

$$\begin{array}{ccc}
\mathrm{Hom}_\Lambda(A', B) & \xrightarrow{\ \ \Delta\ \ } & \mathrm{Ext}_\Lambda^1(A'', B) \\
\downarrow{\scriptstyle \mathrm{Hom}_\Lambda(\gamma_{A'}, B)} & & \downarrow{\scriptstyle \mathrm{Ext}_\Lambda^1(\gamma_{A''}, B)} \\
\mathrm{Hom}_\Lambda(A', B) & \xrightarrow{\ \ \Delta\ \ } & \mathrm{Ext}_\Lambda^1(A'', B)
\end{array}$$

However the vertical maps are multiplications† by γ. Consequently Δ is a Γ-homomorphism.

On the other hand, consider an exact sequence $0 \to B' \to B \to B'' \to 0$. Then

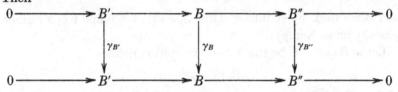

is also a commutative diagram in $\mathscr{C}_\Lambda$ with exact rows. From Theorem 6 we now obtain a commutative diagram

$$\begin{array}{ccc}
\mathrm{Hom}_\Lambda(A, B'') & \xrightarrow{\ \ \Delta\ \ } & \mathrm{Ext}_\Lambda^1(A, B') \\
\downarrow{\scriptstyle \mathrm{Hom}_\Lambda(A, \gamma_{B''})} & & \downarrow{\scriptstyle \mathrm{Ext}_\Lambda^1(A, \gamma_{B'})} \\
\mathrm{Hom}_\Lambda(A, B'') & \xrightarrow{\ \ \Delta\ \ } & \mathrm{Ext}_\Lambda^1(A, B')
\end{array}$$

and the desired result follows as before.

Exercise 8. *Show that if each cyclic left Λ-module is projective, then all left Λ-modules are projective.*

Solution. Suppose that every cyclic left Λ-module is projective. Then Λ/I is projective for every left ideal of Λ. Hence, by Theorem 1, $\mathrm{Ext}_\Lambda^1(\Lambda/I, B) = 0$ for every left Λ-module B and every left ideal I. Theorem 8 now shows that every left Λ-module must be injective. Accordingly, by (Chapter 2, Exercise 14), every left Λ-module is projective.

Exercise 9. *Let $0 \to B \xrightarrow{g} X \xrightarrow{f} A \to 0$ be an exact sequence (in $\mathscr{C}_\Lambda$) and let $\omega : A' \to A$ be a Λ-homomorphism. Show that it is possible to construct*

† See (Chapter 2, Exercise 2) and (Chapter 3, Exercise 2).

(in $\mathscr{C}_\Lambda$) a commutative diagram

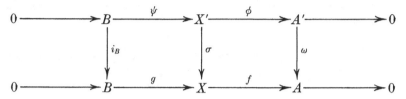

with exact rows.

Solution. Define a Λ-homomorphism $\pi: A' \oplus X \to A$ by

$$\pi(a', x) = -\omega(a') + f(x).$$

Then π is epic because $\pi(0, x) = f(x)$. Now construct an exact sequence

$$0 \to X' \overset{\mu}{\to} A' \oplus X \overset{\pi}{\to} A \to 0 \text{ in } \mathscr{C}_\Lambda.$$

Define the Λ-homomorphism $\phi: X' \to A'$ by $\phi(x') = a'$, where

$$\mu(x') = (a', x).$$

If $a' \in A'$, then we can choose $x \in X$ such that $f(x) = \omega(a')$. In this case $\pi(a', x) = 0$ and hence $(a', x) \in \operatorname{Ker} \pi = \operatorname{Im} \mu$. Consequently there exists $x' \in X'$ such that $\mu(x') = (a', x)$ and therefore $a' = \phi(x')$. Accordingly ϕ is epic.

Now, whenever $b \in B$, $\pi(0, g(b)) = -\omega(0) + f(g(b)) = 0$ and therefore $(0, g(b)) \in \operatorname{Ker} \pi = \operatorname{Im} \mu$. That is $(0, g(b)) = \mu(x')$ for some $x' \in X'$. Define $\psi: B \to X'$ by $\psi(b) = x'$, where $\mu(x') = (0, g(b))$. Then ψ is a well defined Λ-homomorphism since μ is a monomorphism. If

$$\psi(b_1) = \psi(b_2) = x',$$

then $\mu(x') = (0, g(b_1)) = (0, g(b_2))$ and hence $g(b_1) = g(b_2)$. But, since g is monic, this shows that ψ is monic as well.

Let $b \in B$ and put $x' = \psi(b)$. Then $\mu(x') = (0, g(b))$ and therefore $\phi\psi(b) = \phi(x') = 0$. In order to show that the sequence

$$0 \to B \overset{\psi}{\to} X' \overset{\phi}{\to} A' \to 0$$

is exact we need only show that $\operatorname{Ker} \phi \subseteq \operatorname{Im} \psi$. Suppose that $\phi(x') = 0$. Then $\mu(x') = (0, x)$ for some $x \in X$. Now $\pi\mu(x') = 0$. Consequently $f(x) = \pi(0, x) = \pi\mu(x') = 0$. It follows that $x \in \operatorname{Ker} f = \operatorname{Im} g$, that is $x = g(b)$ for some $b \in B$. Hence $\mu(x') = (0, g(b))$ and $x' = \psi(b) \in \operatorname{Im} \psi$.

It remains to obtain a homomorphism $\sigma: X' \to X$ such that $f\sigma = \omega\phi$ and $\sigma\psi = g$. Define $\sigma: X' \to X$ by $\sigma(x') = x$ where $\mu(x') = (a', x)$. Then $f\sigma(x') = f(x)$ and $\omega\phi(x') = \omega(a')$. Now $(a', x) \in \operatorname{Im} \mu = \operatorname{Ker} \pi$ and hence

$0 = \pi(a', x) = -\omega(a') + f(x)$. It now follows that $f\sigma(x') = \omega\phi(x')$. Finally suppose that $b \in B$. Then $\psi(b) = x'$ where $\mu(x') = (0, g(b))$ and therefore $\sigma\psi(b) = g(b)$.

Exercise 10. *Let* $\{A_i\}_{i \in I}$ *be a non-empty family of left* Λ-*modules and* A *their direct sum. Show that* $l.\mathrm{Pd}_\Lambda(A) = \sup_{i \in I} l.\mathrm{Pd}_\Lambda(A_i)$.

Solution. Suppose that $n \geqslant 0$ is an integer and for each $i \in I$ construct an exact sequence

$$0 \to A_{i,n} \to P_{i,n-1} \to \ldots \to P_{i,1} \to P_{i,0} \to A_i \to 0,$$

where each $P_{i,\nu}$ is projective. These sequences give rise, in an obvious manner, to an exact sequence

$$0 \to \bigoplus_{i \in I} A_{i,n} \to \bigoplus_{i \in I} P_{i,n-1} \to \ldots \to \bigoplus_{i \in I} P_{i,0} \to A \to 0$$

and, by (Chapter 2, Theorem 4), $\bigoplus_{i \in I} P_{i,\nu}$ is projective. Again, by (Chapter 2, Theorem 4), $\bigoplus_{i \in I} A_{i,n}$ is projective if and only if $A_{i,n}$ is projective for every i in I. We now see that the following statements are equivalent:

(1) $l.\mathrm{Pd}_\Lambda(A) \leqslant n$;
(2) $\bigoplus_{i \in I} A_{i,n}$ is projective;
(3) $A_{i,n}$ is projective for each $i \in I$;
(4) $l.\mathrm{Pd}_\Lambda(A_i) \leqslant n$ for each $i \in I$;
(5) $\sup_{i \in I} l.\mathrm{Pd}_\Lambda(A_i) \leqslant n$.

The desired result follows.

Exercise 11. *Let* $0 \to A' \xrightarrow{\sigma} A \xrightarrow{\pi} A'' \to 0$ *be an exact sequence of left* Λ-*modules. Establish the following:*

(1) *if* $l.\mathrm{Pd}_\Lambda(A) > l.\mathrm{Pd}_\Lambda(A')$, *then* $l.\mathrm{Pd}_\Lambda(A'') = l.\mathrm{Pd}_\Lambda(A)$;
(2) *if* $l.\mathrm{Pd}_\Lambda(A') > l.\mathrm{Pd}_\Lambda(A)$, *then* $l.\mathrm{Pd}_\Lambda(A'') = l.\mathrm{Pd}_\Lambda(A') + 1$;
(3) *if* $l.\mathrm{Pd}_\Lambda(A') = l.\mathrm{Pd}_\Lambda(A)$, *then* $l.\mathrm{Pd}_\Lambda(A'') \leqslant l.\mathrm{Pd}_A(A) + 1$.

Solution. Construct epimorphisms $\psi':P' \to A'$ and $\psi'':P'' \to A''$ with P' and P'' projective, and put $P = P' \oplus P''$. Then P is also projective. Since π is surjective and P'' is projective, there exists a Λ-homomorphism $\theta:P'' \to A$ such that $\pi\theta = \psi''$. Define

$$\psi:P \to A \quad \text{by} \quad \psi(p', p'') = \sigma\psi'(p') + \theta(p'').$$

Then ψ is a Λ-homomorphism and the diagram

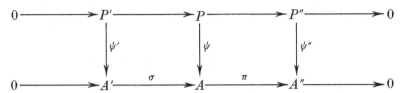

is commutative, where the upper row is formed in the canonical manner. Accordingly if we relabel P', P, P'' as P_0', P_0, P_0'' and use the Ker–Coker sequence, we obtain a commutative diagram

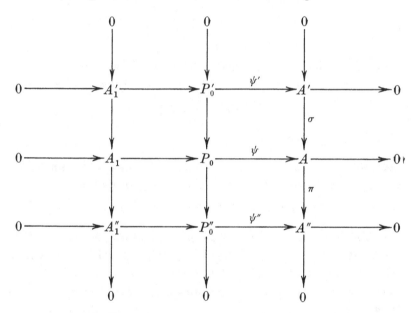

with exact rows and columns.

Put $k = \min(l.\mathrm{Pd}_\Lambda(A'), l.\mathrm{Pd}_\Lambda(A))$. If $k = \infty$, then we are in situation (3) and, in this instance, the assertion is trivial. If $k = -1$, then either $A' = 0$ and $A \neq 0$, or $A' = A = 0$. In both these situations the statement to be proved is obvious. Thus we may suppose that $0 \leqslant k < \infty$. From here on we use induction on k.

Assume that $k = 0$. There are three possibilities namely (a) $A \neq 0$ and projective, but A' is not projective; (b) $A \neq 0$, $A' \neq 0$ and both are projective; (c) $A' \neq 0$ is projective, but A is not projective. Now if A'' is projective, then the sequence $0 \to A' \to A \to A'' \to 0$ splits. It follows that, in case (a), A'' is not projective and $l.\mathrm{Pd}_\Lambda(A'') = l.\mathrm{Pd}_\Lambda(A') + 1$. Case (b) is trivial. Lastly, in case (c), A'' is not projective because otherwise we should have an isomorphism $A \approx A' \oplus A''$ contradicting the assumption that A is not projective. Now in the last diagram we may arrange, on this occasion, that $A'_1 = 0$. Then A_1 and A''_1 are isomorphic and

$$l.\mathrm{Pd}_\Lambda(A) = l.\mathrm{Pd}_\Lambda(A_1) + 1 = l.\mathrm{Pd}_\Lambda(A''_1) + 1 = l.\mathrm{Pd}_\Lambda(A'').$$

This disposes of the case $k = 0$.

We now assume that $1 \leqslant k < \infty$ and make the natural inductive hypothesis. Then $l.\mathrm{Pd}_\Lambda(A'_1) = l.\mathrm{Pd}_\Lambda(A') - 1$ and

$$l.\mathrm{Pd}_\Lambda(A_1) = l.\mathrm{Pd}_\Lambda(A) - 1.$$

If A'' is not projective, then also $l.\mathrm{Pd}_\Lambda(A''_1) = l.\mathrm{Pd}_\Lambda(A'') - 1$ and everything follows by induction. Finally suppose that A'' is projective. It is then possible to arrange that $A''_1 = 0$. This secures that A'_1 and A_1 are isomorphic and hence that $l.\mathrm{Pd}_\Lambda(A') = l.\mathrm{Pd}_\Lambda(A) = k$. Also

$$l.\mathrm{Pd}_\Lambda(A'') < k + 1 = l.\mathrm{Pd}_\Lambda(A) + 1.$$

With this the solution is complete.

Exercise 12. *Let* $\phi: \Lambda \to \Delta$ *be a ring-homomorphism and suppose that* Δ *is a non-trivial ring. Show that if* U *is a left* Δ*-module, then*

$$l.\mathrm{Pd}_\Lambda(U) \leqslant l.\mathrm{Pd}_\Lambda(\Delta) + l.\mathrm{Pd}_\Delta(U).$$

Solution. Put $m = l.\mathrm{Pd}_\Lambda(\Delta)$. Then $m \geqslant 0$ and we may assume that $m < \infty$. We may also assume that $l.\mathrm{Pd}_\Delta(U) < \infty$. The result will be proved by induction on $l.\mathrm{Pd}_\Delta(U)$.

If $l.\mathrm{Pd}_\Delta(U) = -1$, then $U = 0$ and the assertion in this case is trivial.

Next let $l.\mathrm{Pd}_\Delta(U) = 0$. Then U is Δ-projective and hence a direct summand of a free Δ-module $\bigoplus_{i \in I} \Delta$. It follows that U, considered as a Λ-module, is a direct summand of $\bigoplus_{i \in I} \Delta$ also considered as a Λ-module. Since $l.\mathrm{Pd}_\Lambda(\Delta) = m$, Exercise 10 shows that $l.\mathrm{Pd}_\Lambda(\bigoplus_{i \in I} \Delta) = m$. Accordingly

$$l.\mathrm{Pd}_\Lambda(U) \leqslant m = l.\mathrm{Pd}_\Lambda(\Delta) + l.\mathrm{Pd}_\Delta(U).$$

Assume now that $n \geqslant 0$ and that the assertion holds for Δ-modules whose projective dimensions do not exceed n. Let U be a left Δ-module and suppose that $l.\mathrm{Pd}_\Delta(U) = n + 1$. We may construct, in $\mathscr{C}_\Delta$, an exact sequence

$$0 \to V \to F \to U \to 0,$$

where F is Δ-free. Then $l.\mathrm{Pd}_\Lambda(V) = n$ and so, by our induction hypothesis,
$$l.\mathrm{Pd}_\Lambda(V) \leqslant m + l.\mathrm{Pd}_\Delta(V) \leqslant m + n.$$

But $0 \to V \to F \to U \to 0$ is an exact sequence of Λ-modules and our previous remarks show that $l.\mathrm{Pd}_\Lambda(F) \leqslant m$. Accordingly, by Exercise 11,
$$l.\mathrm{Pd}_\Lambda(U) \leqslant m + n + 1 = l.\mathrm{Pd}_\Lambda(\Delta) + l.\mathrm{Pd}_\Delta(U).$$

This completes the solution.

Exercise 13. *Determine* $\mathrm{Pd}_Z(Q)$ *and* $\mathrm{Id}_Z(Z)$, *where* Q *is the field of rational numbers.*

Solution. We can construct an exact sequence $0 \to A \to F \to Q \to 0$ in $\mathscr{C}_Z$, where F is free and A is a submodule of F. By (Chapter 2, Exercise 6), A is free and hence projective. Now Q is not projective. (For if it were then, by (Chapter 2, Exercise 6), it would be free as well as divisible. However no non-zero free Z-module is divisible.) It follows that $\mathrm{Pd}_Z(Q) = 1$.

Consider the exact sequence
$$0 \to Z \to Q \to Q/Z \to 0.$$

Both Q and Q/Z are divisible and therefore injective Z-modules by (Chapter 2, Theorem 12). On the other hand, Z is not divisible and therefore not injective. It follows that $\mathrm{Id}_Z(Z) = 1$.

Exercise 14. *Let* $\{B_i\}_{i \in I}$ *be a family of left* Λ-*modules and let* B *be their direct product. Show that*
$$l.\mathrm{Id}_\Lambda(B) = \sup_{i \in I} l.\mathrm{Id}_\Lambda(B_i).$$

Solution. The argument is a simple adaptation of that used to solve Exercise 10.

Exercise 15. *Show that* $\mathrm{GD}(Z) = 1$.

Solution. For every ideal I, of Z, we have an exact sequence
$$0 \to I \to Z \to Z/I \to 0.$$

Now Z is projective and, if $I \neq 0$, I and Z are isomorphic Z-modules. Consequently I is projective. It follows that $\mathrm{Pd}_Z(Z/I) \leqslant 1$. Hence, by Theorem 18, $\mathrm{GD}(Z) \leqslant 1$. But, by Exercise 13, $\mathrm{Pd}_Z(Q) = 1$. It follows that $\mathrm{GD}(Z) = 1$.

Exercise 16. *Suppose that*
$$0 \to B \to E_0 \to E_1 \to E_2 \to \ldots$$

*is a minimal injective resolution of the left Λ-module B and let γ be an
element of the centre of Λ which is not a zerodivisor on either Λ or B.
Show that if $D = \mathrm{Im}(E_0 \to E_1)$, then*

$$0 \to \mathrm{Hom}_\Lambda(\Lambda/\gamma\Lambda, D) \to \mathrm{Hom}_\Lambda(\Lambda/\gamma\Lambda, E_1) \to \mathrm{Hom}_\Lambda(\Lambda/\gamma\Lambda, E_2) \to \ldots$$

*is a minimal injective resolution of the $\Lambda/\gamma\Lambda$-module $\mathrm{Hom}_\Lambda(\Lambda/\gamma\Lambda, D)$.
Show also that $\mathrm{Hom}_\Lambda(\Lambda/\gamma\Lambda, D)$ and $B/\gamma B$ are isomorphic $\Lambda/\gamma\Lambda$-
modules.*

Solution. Let E be an injective Λ-module. Since multiplication by γ
induces a monomorphism $\Lambda \to \Lambda$, the resulting homomorphism

$$\mathrm{Hom}_\Lambda(\Lambda, E) \to \mathrm{Hom}_\Lambda(\Lambda, E)$$

is epic. Accordingly $\mathrm{Hom}_\Lambda^{\cdot}(\Lambda, E) = \gamma\,\mathrm{Hom}_\Lambda(\Lambda, E)$. However, by
(Chapter 2, Example 1), $\mathrm{Hom}_\Lambda(\Lambda, E)$ and E are isomorphic Λ-modules.
Consequently $E = \gamma E$.

Let C be a Λ-module and $0:_C\gamma$ the submodule annihilated by γ.
Then, as we saw in the discussion of (Chapter 2, Exercise 20), we have
a $\Lambda/\gamma\Lambda$-isomorphism $\mathrm{Hom}_\Lambda(\Lambda/\gamma\Lambda, C) \approx 0:_C\gamma$ which is natural for Λ-
homomorphisms $f:C \to C^*$.

Let $0 \to C' \to C \to C'' \to 0$ be exact in $\mathscr{C}_\Lambda$ and consider the commutative diagram

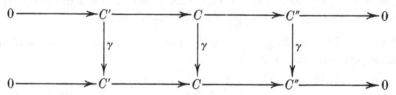

The theory of the Ker–Coker sequence shows that we have an exact
sequence $$0 \to (0:_{C'}\gamma) \to (0:_C\gamma) \to (0:_{C''}\gamma) \to C'/\gamma C'.$$

Accordingly if $C' = \gamma C'$, then

$$0 \to \mathrm{Hom}_\Lambda(\Lambda/\gamma\Lambda, C') \to \mathrm{Hom}_\Lambda(\Lambda/\gamma\Lambda, C) \to \mathrm{Hom}_\Lambda(\Lambda/\gamma\Lambda, C'') \to 0$$

is exact in $\mathscr{C}_{\Lambda/\gamma\Lambda}$.
Put $D_0 = D$ and $D_i = \mathrm{Im}(E_i \to E_{i+1})$ for $i \geqslant 1$. Then D_j is the
image of the injective Λ-module E_j and we know that $E_j = \gamma E_j$. Thus
$0 \to D_j \to E_{j+1} \to D_{j+1} \to 0$ is exact and $D_j = \gamma D_j$. Accordingly

$$0 \to \mathrm{Hom}_\Lambda(\Lambda/\gamma\Lambda, D_j) \to \mathrm{Hom}_\Lambda(\Lambda/\gamma\Lambda, E_{j+1}) \to \mathrm{Hom}_\Lambda(\Lambda/\gamma\Lambda, D_{j+1}) \to 0$$

is exact in $\mathscr{C}_{\Lambda/\gamma\Lambda}$. Further $D_j \to E_{j+1}$ is an essential monomorphism and E_{j+1} is Λ-injective. It follows, from (Chapter 2, Exercise 20), that $\mathrm{Hom}_\Lambda (\Lambda/\gamma\Lambda, D_j) \to \mathrm{Hom}_\Lambda (\Lambda/\gamma\Lambda, E_{j+1})$ is an essential monomorphism in $\mathscr{C}_{\Lambda/\gamma\Lambda}$ and $\mathrm{Hom}_\Lambda (\Lambda/\gamma\Lambda, E_{j+1})$ is $\Lambda/\gamma\Lambda$-injective. Accordingly

$$0 \to \mathrm{Hom}_\Lambda (\Lambda/\gamma\Lambda, D) \to \mathrm{Hom}_\Lambda (\Lambda/\gamma\Lambda, E_1) \to \mathrm{Hom}_\Lambda (\Lambda/\gamma\Lambda, E_2) \to \ldots$$

is a minimal injective resolution of the $\Lambda/\gamma\Lambda$-module $\mathrm{Hom}_\Lambda (\Lambda/\gamma\Lambda, D)$.

We have $B \cap \mathrm{Ker}\, (E_0 \overset{\gamma}{\to} E_0) = \mathrm{Ker}\, (B \overset{\gamma}{\to} B) = 0$ because γ is not a zerodivisor on B. It follows that $\mathrm{Ker}\, (E_0 \overset{\gamma}{\to} E_0) = 0$ because E_0 is an essential extension of B. But $\gamma E_0 = E_0$ because E_0 is injective. Thus multiplication by γ induces an isomorphism on E_0. Consider the commutative diagram

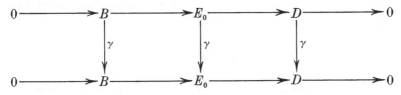

Since this has exact rows, the theory of the Ker–Coker sequence shows that we have a $\Lambda/\gamma\Lambda$-isomorphism $0 :_D \gamma \approx B/\gamma B$. It follows that $\mathrm{Hom}_\Lambda (\Lambda/\gamma\Lambda, D)$ and $B/\gamma B$ are isomorphic in $\mathscr{C}_{\Lambda/\gamma\Lambda}$ and this completes the solution.

Exercise 17. *Let B be a left Λ-module and γ an element in the centre of Λ which is not a zerodivisor on either Λ or B. Suppose that $l.\mathrm{Id}_\Lambda (B) < n$, where $1 \leqslant n < \infty$. Show that $l.\mathrm{Id}_{\Lambda/\gamma\Lambda} (B/\gamma B) < n - 1$.*

Solution. Let $0 \to B \to E_0 \to E_1 \to \ldots$ be a minimal injective resolution of B. By Exercise 16,

$$0 \to \mathrm{Hom}_\Lambda (\Lambda/\gamma\Lambda, D) \to \mathrm{Hom}_\Lambda (\Lambda/\gamma\Lambda, E_1) \to \mathrm{Hom}_\Lambda (\Lambda/\gamma\Lambda, E_2) \to \ldots,$$

where $D = \mathrm{Im}\, (E_0 \to E_1)$, is a minimal injective resolution of the $\Lambda/\gamma\Lambda$-module $\mathrm{Hom}_\Lambda (\Lambda/\gamma\Lambda, D)$. By Theorem 20, $E_i = 0$ for all $i \geqslant n$ and hence $\mathrm{Hom}_\Lambda (\Lambda/\gamma\Lambda, E_i) = 0$ for all $i \geqslant n$. Consequently

$$l.\mathrm{Id}_{\Lambda/\gamma\Lambda} (\mathrm{Hom}_\Lambda (\Lambda/\gamma\Lambda, D)) < n - 1.$$

Exercise 16 also shows that $\mathrm{Hom}_\Lambda (\Lambda/\gamma\Lambda, D)$ and $B/\gamma B$ are isomorphic $\Lambda/\gamma\Lambda$-modules. Accordingly $l.\mathrm{Id}_{\Lambda/\gamma\Lambda} (B/\gamma B) < n - 1$ as required.

Exercise 18. *Let I be a two-sided ideal of Λ and P a projective left Λ-module. Show that P/IP is a projective left Λ/I-module.*

Solution. Since P is projective there is a free Λ-module F and Λ-homomorphisms $\sigma:P\to F$ and $\pi:F\to P$ such that $\pi\sigma = i_P$. Let $\{e_j\}_{j\in J}$ be a base for F and let $\bar{e}_j$ be the natural image of e_j in F/IF. An easy verification shows that $\{\bar{e}_j\}_{j\in J}$ is a base for F/IF as a module over Λ/I, so that, in particular, F/IF is Λ/I-free.

The homomorphisms σ and π give rise to Λ/I-homomorphisms $\bar{\sigma}:P/IP\to F/IF$ and $\bar{\pi}:F/IF\to P/IP$ and their composition is an identity map. It follows that $\bar{\sigma}$ is a monomorphism and $\mathrm{Im}\,\bar{\sigma}$ is a direct summand of F/IF and therefore Λ/I-projective. But P/IP and $\mathrm{Im}\,\bar{\sigma}$ are isomorphic. Consequently P/IP is Λ/I-projective as well.

Exercise 19. *Suppose that* $0\to B\overset{\psi}{\to} X\overset{\phi}{\to} A\to 0$ *and* $0\to B\overset{\psi'}{\to} X'\overset{\phi'}{\to} A\to 0$ *are exact sequences, and* $f:X\to X'$ *is a Λ-homomorphism such that the diagram*

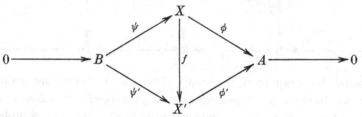

is commutative. Show that f is an isomorphism.

Solution. Suppose that $f(x) = 0$. Then $\phi(x) = \phi'f(x) = 0$ and hence $x = \psi(b)$ for some $b\in B$. But $\psi'(b) = f\psi(b) = 0$ and, since ψ' is a monomorphism, $b = 0$. Thus $x = 0$ and therefore f is a monomorphism.

Now assume that $x'\in X'$. Since ϕ is an epimorphism, there exists $x\in X$ such that $\phi(x) = \phi'(x')$. Then $\phi'(x' - f(x)) = \phi'(x') - \phi(x) = 0$. Hence $x' - f(x) = \psi'(b)$ for some $b\in B$. It follows that

$$f(\psi(b) + x) = \psi'(b) + f(x) = x'.$$

Thus f is also an epimorphism and hence an isomorphism.

Exercise 20. *Show that if S is an extension of B by A, then*

$$\Delta^1(S) = \Delta^2(S).$$

Solution. Let S be the exact sequence $0\to B\overset{\psi}{\to} X\overset{\phi}{\to} A\to 0$. We can construct exact sequences $0\to A_1\overset{\sigma}{\to} P\overset{\pi}{\to} A\to 0$ and

$$0\to B\overset{\sigma'}{\to} E\overset{\pi'}{\to} B_1\to 0,$$

where P is projective and E injective. Hence we obtain commutative diagrams

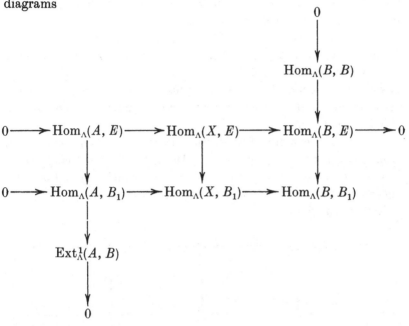

and

with exact rows and columns. Further $\Delta^1(S)$ resp. $\Delta^2(S)$ is, by definition, the image of i_B resp. i_A under the connecting homomorphism arising from the first resp. second diagram.

Consider the first diagram. The image of i_B in $\mathrm{Hom}_\Lambda\,(B,E)$ is σ'. Choose $f \in \mathrm{Hom}_\Lambda\,(X,E)$ to have image σ' in $\mathrm{Hom}_\Lambda\,(B,E)$, i.e. so that $f\psi = \sigma'$. The image of f in $\mathrm{Hom}_\Lambda\,(X,B_1)$ is $\pi'f$. Accordingly, by the theory of the connecting homomorphism (see section (3.4)), there exists $g \in \mathrm{Hom}_\Lambda\,(A,B_1)$ such that $g\phi = \pi'f$. Then $\Delta^1(S)$ is the image $\bar{g}$, of g, in $\mathrm{Ext}^1_\Lambda\,(A,B)$.

Let us now examine the second diagram. The image of i_A in $\mathrm{Hom}_\Lambda\,(P,A)$ is π and there exists $f' \in \mathrm{Hom}_\Lambda\,(P,X)$ such that $\phi f' = \pi$. Further there exists $g' \in \mathrm{Hom}_\Lambda\,(A_1,B)$ for which $\psi g' = f'\sigma$. This ensures that $\Delta^2(S)$ is the image $\bar{g}'$, of g', in $\mathrm{Ext}^1_\Lambda\,(A,B)$.

Next the diagram

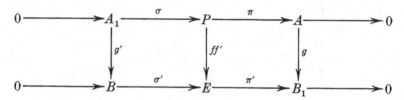

is commutative because $ff'\sigma = f\psi g' = \sigma'g'$ and $\pi'ff' = g\phi f' = g\pi$. Hence g' and g are associated in the sense of section (3.2). It follows that $\bar{g}' = \bar{g}$ and therefore $\Delta^2(S) = \Delta^1(S)$.

Exercise 21. *Let $S_1 : 0 \to B \xrightarrow{\psi_1} X_1 \xrightarrow{\phi_1} A \to 0$ and $S_2 : 0 \to B \xrightarrow{\psi_2} X_2 \xrightarrow{\phi_2} A \to 0$ be two extensions of B by A and suppose that they correspond to the elements ω_1 and ω_2 of $\mathrm{Ext}^1_\Lambda\,(A,B)$. Construct an extension of B by A that corresponds to $\omega_1 + \omega_2$.*

Solution. Let Y be the submodule of $X_1 \oplus X_2$ consisting of all pairs (x_1, x_2) such that $\phi_1(x_1) = \phi_2(x_2)$, and let Y' be the submodule of $X_1 \oplus X_2$ formed by all pairs $(-\psi_1(b), \psi_2(b))$, where $b \in B$. Then Y' is a submodule of Y. Put $X = Y/Y'$ and, if $(x_1, x_2) \in Y$, let $\overline{(x_1, x_2)}$ denote its natural image in X. Define a Λ-homomorphism $\psi : B \to X$ by $\psi(b) = \overline{(\psi_1(b), 0)} = \overline{(0, \psi_2(b))}$. If now $\psi(b) = 0$, then

$$(\psi_1(b), 0) = (-\psi_1(b'), \psi_2(b'))$$

for some $b' \in B$. Thus $\psi_2(b') = 0$ and, since ψ_2 is a monomorphism, $b' = 0$. It follows that $\psi_1(b) = 0$ and hence that $b = 0$. This shows that ψ is a monomorphism.

Define a Λ-homomorphism $\phi: X \to A$ by $\phi(\overline{(x_1, x_2)}) = \phi_1(x_1) = \phi_2(x_2)$. It is easily verified that ϕ is well defined. Moreover, if $a \in A$, then there exist $x_1 \in X_1$ and $x_2 \in X_2$ such that $\phi_1(x_1) = \phi_2(x_2) = a$. Hence $a = \phi(\overline{(x_1, x_2)})$. This shows that ϕ is an epimorphism.

We claim that

$$(S) \quad 0 \to B \xrightarrow{\psi} X \xrightarrow{\phi} A \to 0$$

is an exact sequence and therefore it provides an extension of B by A.

For suppose that $\phi(\overline{(x_1, x_2)}) = 0$. Then $x_1 = \psi_1(b_1)$ and $x_2 = \psi_2(b_2)$ for suitable elements b_1, b_2 in B. Accordingly

$$\overline{(x_1, x_2)} = \overline{(\psi_1(b_1), 0)} + \overline{(0, \psi_2(b_2))} = \psi(b_1 + b_2).$$

It follows that $\mathrm{Ker}\, \phi \subseteq \mathrm{Im}\, \psi$ and, as the opposite inclusion is obvious, this establishes our claim.

Let $0 \to A_1 \xrightarrow{\beta} P \xrightarrow{\alpha} A \to 0$ be an exact sequence in $\mathscr{C}_\Lambda$ with P projective and form commutative diagrams

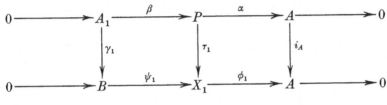

and

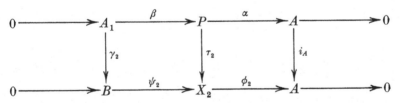

Next define a Λ-homomorphism $\tau: P \to X$ by $\tau(p) = \overline{(\tau_1(p), \tau_2(p))}$. This is possible because $\phi_1\tau_1(p) = \phi_2\tau_2(p) = \alpha(p)$. A straightforward verification shows that the diagram

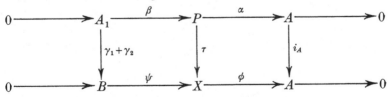

is commutative.

Finally let $\Omega:\mathrm{Hom}_\Lambda\,(A_1, B) \to \mathrm{Ext}^1_\Lambda\,(A, B)$ be the connecting homomorphism arising from the sequence $0 \to A_1 \to P \to A \to 0$. By (3.9.9), $\Omega(\gamma_1) = \Delta(S_1) = \omega_1$ and likewise $\Omega(\gamma_2) = \omega_2$. It follows that

$$\Delta(S) = \Omega(\gamma_1 + \gamma_2) = \Omega(\gamma_1) + \Omega(\gamma_2) = \omega_1 + \omega_2$$

which shows that S is the required extension.

4

POLYNOMIAL RINGS AND
MATRIX RINGS

4.1 General

As usual Λ, Γ, Δ denote rings which have identity elements (but which need not be commutative) and the symbols $\mathscr{C}_\Lambda, \mathscr{C}_\Lambda^L, \mathscr{C}_\Lambda^R$ etc. retain the significance given to them in section (1.1). We shall continue to use Z to denote the ring of integers and, when convenient, identify the category of additive abelian groups with the category of Z-modules. Throughout section (4.2), the letter x will be used for an indeterminate. We recall that a *non-trivial* ring is one in which the zero element is different from the identity element.

4.2 The polynomial functor

Let x be an indeterminate and let us denote by $\Lambda[x]$ the set of *polynomials* in x with coefficients in Λ. Thus if f and g belong to $\Lambda[x]$, then they are expressions of the form

$$f = \lambda_0 + \lambda_1 x + \lambda_2 x^2 + \ldots + \lambda_m x^m$$

and

$$g = \lambda_0' + \lambda_1' x + \lambda_2' x^2 + \ldots + \lambda_n' x^n,$$

where λ_i and λ_j' belong to Λ. Let us define the *sum* and *product* of f and g by

$$f + g = (\lambda_0 + \lambda_0') + (\lambda_1 + \lambda_1') x + (\lambda_2 + \lambda_2') x^2 + \ldots$$

and $\quad fg = (\lambda_0 \lambda_0') + (\lambda_0 \lambda_1' + \lambda_1 \lambda_0') x + (\lambda_0 \lambda_2' + \lambda_1 \lambda_1' + \lambda_2 \lambda_0') x^2 + \ldots,$

then the result is that $\Lambda[x]$ becomes a ring having the constant polynomials $0 + 0x + 0x^2 + \ldots$ and $1 + 0x + 0x^2 + \ldots$ as its zero and identity element respectively.

The indeterminate x may be identified with the polynomial

$$0 + 1x + 0x^2 + 0x^3 + \ldots$$

and if this is done then x belongs to the centre of $\Lambda[x]$ and is not a zerodivisor. Note that if Λ is non-trivial, then x is also a non-unit of $\Lambda[x]$.

Consider the mapping $\Lambda[x] \to \Lambda$ in which each polynomial is mapped on to its constant term. This is a surjective ring-homomorphism whose kernel is the (two-sided) ideal $(x) = x\Lambda[x]$ generated by x. It follows that $\Lambda[x]/(x)$ and Λ are isomorphic rings. We normally identify them and write

$$\Lambda[x]/(x) = \Lambda. \qquad (4.2.1)$$

Let A be a left Λ-module. We may also consider formal polynomials in x with *coefficients in A*. The set of all such polynomials is denoted by $A[x]$. Thus each element ξ of $A[x]$ is an expression of the form $\xi = a_0 + a_1 x + a_2 x^2 + \ldots + a_s x^s$, where a_0, a_1, a_2, etc. belong to A. It is clear how members of $A[x]$ are to be added and it is also clear that, once addition has been defined, $A[x]$ constitutes an abelian group.

Suppose that $f = \lambda_0 + \lambda_1 x + \ldots + \lambda_m x^m$ and $\xi = a_0 + a_1 x + \ldots + a_s x^s$ belong to $\Lambda[x]$ and $A[x]$ respectively. We define the product $f\xi$ by

$$f\xi = (\lambda_0 a_0) + (\lambda_0 a_1 + \lambda_1 a_0) x + (\lambda_0 a_2 + \lambda_1 a_1 + \lambda_2 a_0) x^2 + \ldots$$

and it is now apparent that $A[x]$ has taken on the structure of a left $\Lambda[x]$-module. However we can say more. Suppose that $u : A \to B$ is a homomorphism of left Λ-modules. This will induce a $\Lambda[x]$-homomorphism $A[x] \to B[x]$ in which $a_0 + a_1 x + a_2 x^2 + \ldots$ in $A[x]$ is mapped into the element $u(a_0) + u(a_1) x + u(a_2) x^2 + \ldots$ of $B[x]$. In fact we have constructed a covariant functor from $\mathscr{C}_\Lambda^L$ to $\mathscr{C}_{\Lambda[x]}^L$. This will be called the *polynomial functor*.

Exercise 1. *Show that the polynomial functor is exact. Show also that if the left Λ-module A is free, then $A[x]$ is a free $\Lambda[x]$-module. Deduce that a left Λ-module B is Λ-projective if and only if $B[x]$ is $\Lambda[x]$-projective.*

As an application we shall prove

Lemma 1. *Let A be a left Λ-module. Then*

$$l.\mathrm{Pd}_\Lambda (A) = l.\mathrm{Pd}_{\Lambda[x]} (A[x]).$$

Proof. Let $n \geqslant 0$ be an integer and let us construct an exact sequence

$$0 \to A_n \to P_{n-1} \to \ldots \to P_1 \to P_0 \to A \to 0,$$

where each P_i is Λ-projective. Applying the polynomial functor we find, using Exercise 1, that

$$0 \to A_n[x] \to P_{n-1}[x] \to \ldots \to P_1[x] \to P_0[x] \to A[x] \to 0$$

is an exact sequence of $\Lambda[x]$-modules in which each $P_i[x]$ is $\Lambda[x]$-projective. Next $l.\mathrm{Pd}_\Lambda (A) \leqslant n$ if and only if A_n is projective, whereas

$l.\mathrm{Pd}_{\Lambda[x]}(A[x]) \leqslant n$ if and only if $A_n[x]$ is $\Lambda[x]$-projective. However, by Exercise 1, for A_n to be Λ-projective it is necessary and sufficient that $A_n[x]$ be $\Lambda[x]$-projective. Thus $l.\mathrm{Pd}_{\Lambda}(A) \leqslant n$ is equivalent to $l.\mathrm{Pd}_{\Lambda[x]}(A[x]) \leqslant n$ and now the lemma follows.

We recall that $l.\mathrm{GD}(\Lambda)$ denotes the left global dimension† of Λ.

Theorem 1. *If Λ is a non-trivial ring, then*

$$l.\mathrm{GD}(\Lambda[x]) = l.\mathrm{GD}(\Lambda) + 1.$$

Proof. If $l.\mathrm{GD}(\Lambda) = \infty$, then $l.\mathrm{GD}(\Lambda[x]) = \infty$ as well by virtue of Lemma 1. We may therefore suppose that $l.\mathrm{GD}(\Lambda) = n$, where $0 \leqslant n < \infty$. In this case there exists a Λ-module A such that

$$l.\mathrm{Pd}_{\Lambda}(A) = n.$$

It follows, from (4.2.1), that $l.\mathrm{Pd}_{\Lambda[x]/(x)}(A) = n$ and now, by (Chapter 3, Theorem 21), we have $l.\mathrm{Pd}_{\Lambda[x]}(A) = n+1$. Accordingly

$$l.\mathrm{GD}(\Lambda[x]) \geqslant n+1.$$

To complete the proof we shall assume that M is a left $\Lambda[x]$-module and show that $l.\mathrm{Pd}_{\Lambda[x]}(M) \leqslant n+1$. To this end let N consist of all sequences $\{m_0, m_1, m_2, \ldots\}$, where $m_i \in M$ and almost all the m_i are zero. Then N becomes an abelian group if we add two such sequences term by term. We can now turn N into a left $\Lambda[x]$-module in such a way that

$$(\lambda x^{\nu})\{m_0, m_1, m_2, \ldots\} = \{\mu_0, \mu_1, \mu_2, \ldots\}$$

where $\mu_j = 0$ if $j < \nu$ and $\mu_i = \lambda m_{i-\nu}$ if $i \geqslant \nu$. In order to identify this module, we observe that the mapping $\Lambda \to \Lambda[x]$ which takes $\lambda \in \Lambda$ into the corresponding constant polynomial is a ring-homomorphism. Let us use it to regard M as a Λ-module. If we do this and then form $M[x]$, it is clear that N and $M[x]$ are isomorphic $\Lambda[x]$-modules under an isomorphism in which $\{m_0, m_1, m_2, \ldots\}$ corresponds to the polynomial with the m_i as coefficients. It follows that

$$l.\mathrm{Pd}_{\Lambda[x]}(N) = l.\mathrm{Pd}_{\Lambda[x]}(M[x]) = l.\mathrm{Pd}_{\Lambda}(M)$$

by Lemma 1. Consequently $l.\mathrm{Pd}_{\Lambda[x]}(N) \leqslant n$.

We next define a mapping $\phi: N \to M$ by requiring that the image of $\{m_0, m_1, m_2, \ldots\}$ be $\sum_i x^i m_i$. (Remember that M is a $\Lambda[x]$-module.) Then ϕ is surjective and an easy verification shows it to be a $\Lambda[x]$-homomorphism.

† See section (3.7).

Suppose now that $\{m_0, m_1, m_2, ...\}$ belongs to $\mathrm{Ker}\,\phi$. Then $\sum_i x^i m_i = 0$.

Put
$$\mu_0 = -(m_1 + xm_2 + x^2 m_3 + ...),$$
$$\mu_1 = -(m_2 + xm_3 + x^2 m_4 + ...),$$
$$\mu_2 = -(m_3 + xm_4 + x^2 m_5 + ...)$$

and so on. It follows that $\{\mu_0, \mu_1, \mu_2, ...\}$ belongs to N and also that

$$\{m_0, m_1, m_2, ...\} = \{x\mu_0, x\mu_1 - \mu_0, x\mu_2 - \mu_1, ...\}.$$

We can, in fact, obtain a mapping $N \to \mathrm{Ker}\,\phi$ by associating the element $\{\mu_0, \mu_1, \mu_2, ...\}$ of N with the element

$$\{x\mu_0, x\mu_1 - \mu_0, x\mu_2 - \mu_1, ...\}$$

of $\mathrm{Ker}\,\phi$. This particular mapping is compatible with addition and we have just proved that it is surjective. Also if $\mu_s \neq 0$ and $\mu_j = 0$ for all $j > s$, then $x\mu_{s+1} - \mu_s \neq 0$. It follows that $N \to \mathrm{Ker}\,\phi$ is an isomorphism of abelian groups and an easy verification shows that it is indeed an isomorphism of $\Lambda[x]$-modules. Thus we have an exact sequence
$$0 \to N \xrightarrow{\phi} N \to M \to 0$$

of $\Lambda[x]$-modules. Accordingly, by (Chapter 3, Exercise 11),

$$l.\mathrm{Pd}_{\Lambda[x]}(M) \leqslant l.\mathrm{Pd}_{\Lambda[x]}(N) + 1.$$

Since we have already shown that $l.\mathrm{Pd}_{\Lambda[x]}(N) \leqslant n$, this completes the proof.

This result is readily generalized. Let $x_1, x_2, ..., x_s$ be distinct indeterminates. Then we can form, in an obvious manner, the ring $\Lambda[x_1, x_2, ..., x_s]$ of polynomials in $x_1, x_2, ..., x_s$ with coefficients in Λ. Here it is understood that the various indeterminates commute with one another and also with the elements of the ground-ring Λ. Thus $x_1, x_2, ..., x_s$ are all in the centre of $\Lambda[x_1, x_2, ..., x_s]$. Suppose, for the moment, that $s \geqslant 2$ and put $\Lambda^* = \Lambda[x_1, x_2, ..., x_{s-1}]$. Then every member of $\Lambda[x_1, x_2, ..., x_s]$ may be regarded as a polynomial in x_s with coefficients in Λ^* so that we have, in fact, a ring-identity

$$\Lambda[x_1, x_2, ..., x_{s-1}, x_s] = \Lambda^*[x_s].$$

In view of this Theorem 1 yields, by a straightforward induction on the number of variables, the following more general result.

Theorem 2. *Let Λ be a non-trivial ring and let $x_1, x_2, ..., x_s$ be indeterminates. Then*
$$l.\mathrm{GD}\,(\Lambda[x_1, x_2, ..., x_s]) = l.\mathrm{GD}\,(\Lambda) + s.$$

Exercise 2. *Let Λ be a non-trivial ring and $x_1, x_2, \ldots, x_s$ indeterminates. Put $\Lambda[x] = \Lambda[x_1, x_2, \ldots, x_s]$. Show that*

$$l. \operatorname{Pd}_{\Lambda[x]}\{\Lambda[x]/(x_1\Lambda[x] + x_2\Lambda[x] + \ldots + x_s\Lambda[x])\} = s.$$

There is a special case of Theorem 2 which, for historical reasons, merits a special mention. This is

Theorem 3. *Let K be a field and $x_1, x_2, \ldots, x_s$ indeterminates. Then* $\operatorname{GD}(K[x_1, x_2, \ldots, x_s]) = s.$

Remark. This result is known as *Hilbert's Syzygies Theorem*.

Proof. By Theorem 2 it is enough to show that $\operatorname{GD}(K) = 0$, i.e. that every K-module is projective. But a K-module is just a vector space over K and as such it possesses a base. Accordingly every K-module is free and with this the proof is complete.

4.3 Generators of a category

The ring Λ can be regarded as a left module with respect to itself, that is to say it can be regarded as an object of the category $\mathscr{C}_\Lambda^L$. Of course, it is a very special object. In terms of a concept to be introduced shortly, its important properties are described by saying that Λ is a *generator* of the category. However there are other modules which share these properties. We shall characterize the typical generator K by means of the functor $\operatorname{Hom}_\Lambda(K, -)$ to which it gives rise.

To this end let $F: \mathscr{C}_\Lambda \to \mathscr{C}_\Delta$ be an *additive* covariant functor and suppose that C, D are Λ-modules. If now $f \in \operatorname{Hom}_\Lambda(C, D)$ then $F(f)$ belongs to $\operatorname{Hom}_\Lambda(F(C), F(D))$ and the association of $F(f)$ with f provides a homomorphism

$$\operatorname{Hom}_\Lambda(C, D) \to \operatorname{Hom}_\Lambda(F(C), F(D)) \tag{4.3.1}$$

of abelian groups.

Definition. *The additive covariant functor $F: \mathscr{C}_\Lambda \to \mathscr{C}_\Delta$ is said to be 'faithful' if (4.3.1) is a monomorphism for all choices of C and D.*

Let K be a Λ-module. Then $\operatorname{Hom}_\Lambda(K, -)$ is an additive covariant functor from $\mathscr{C}_\Lambda$ to $\mathscr{C}_Z$.

Definition. *The Λ-module K is said to be a 'generator' of the category $\mathscr{C}_\Lambda$ if the functor $\operatorname{Hom}_\Lambda(K, -)$, from $\mathscr{C}_\Lambda$ to $\mathscr{C}_Z$, is faithful.*

Theorem 4. *Let K be a generator of $\mathscr{C}_\Lambda$ and A a Λ-module. Then A is the sum of its submodules $f(K)$, where f ranges over $\operatorname{Hom}_\Lambda(K, A)$.*

110 POLYNOMIAL RINGS AND MATRIX RINGS

Proof. Let B be the sum of the submodules $f(K)$ and let $\phi : A \to A/B$ be the canonical homomorphism. Put $F = \mathrm{Hom}_\Lambda (K, -)$. Then

$$F(\phi) : \mathrm{Hom}_\Lambda (K, A) \to \mathrm{Hom}_\Lambda (K, A/B).$$

Indeed if $f \in \mathrm{Hom}_\Lambda (K, A)$, then

$$(F(\phi))(f) = \phi f = 0$$

because $f(K) \subseteq B = \mathrm{Ker}\, \phi$. Thus $F(\phi) = 0$ and therefore $\phi = 0$ because F is faithful. It follows that $B = A$ as required.

Theorem 5. *Let K be a generator of $\mathscr{C}_\Lambda$ and A a Λ-module. Then A is a homomorphic image of a direct sum of copies of K.*

Proof. We define a module C by

$$C = \sum_{\mathrm{Hom}_\Lambda(K, A)} K \quad \text{(direct sum)},$$

i.e. C is a direct sum in which each summand is K and there is one summand for every member of $\mathrm{Hom}_\Lambda (K, A)$. A typical element of C is a family $\{\alpha_f\}_{f \in \mathrm{Hom}_\Lambda(K, A)}$, where $\alpha_f \in K$ and almost all α_f are zero.

Let $\psi : C \to A$ be the Λ-homomorphism in which the family $\{\alpha_f\}$ is mapped into $\sum_f f(\alpha_f)$. Obviously, for each f in $\mathrm{Hom}_\Lambda (K, A)$, $\psi(C)$ contains $f(K)$. It follows, from Theorem 4, that $\psi(C) = A$. This completes the proof.

We are now in a position to describe the typical generator of $\mathscr{C}_\Lambda$ in module-theoretic terms.

Theorem 6. *Let Λ be a non-trivial ring and let K be a module in $\mathscr{C}_\Lambda$. Then K is a generator of $\mathscr{C}_\Lambda$ if and only if there is a direct sum of copies of K which has a non-zero free module as a direct summand.*

Proof. Suppose first that K is a generator. By Theorem 5, we can construct an exact sequence $0 \to B \to C \to \Lambda \to 0$ in $\mathscr{C}_\Lambda$, where C is a direct sum of copies of K. But Λ is projective and therefore, by (Chapter 2, Theorem 5), the sequence splits. Consequently Λ is isomorphic to a direct summand of C.

Now assume that C is a direct sum of copies of K, say $C = \bigoplus_I K$, and that $\Phi \neq 0$ is a free submodule of C which is also a direct summand of that module. Evidently we may suppose that Φ has a base consisting of a single element e (say).

Let A be a Λ-module. By (Chapter 2, Theorem 2) we have an iso-morphism
$$\operatorname{Hom}_\Lambda(C, A) \xrightarrow{\sim} \prod_I \operatorname{Hom}_\Lambda(K, A)$$
of abelian groups in which f in $\operatorname{Hom}_\Lambda(C, A)$ is mapped into $\{f\sigma_i\}_{i \in I}$, where $\sigma_i : K \to C$ is the typical embedding homomorphism. Further if $u : A \to A'$ is a Λ-homomorphism, then the diagram

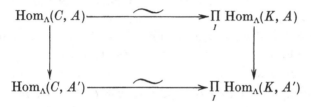

is commutative. Here the right vertical homomorphism is induced by u on a term by term basis.

Assume that $u : A \to A'$ is such that the induced homomorphism $\operatorname{Hom}_\Lambda(K, A) \to \operatorname{Hom}_\Lambda(K, A')$ is null. We wish to show that u itself is a null homomorphism. Now the above remarks show that
$$\operatorname{Hom}_\Lambda(C, A) \to \operatorname{Hom}_\Lambda(C, A')$$
is null, that is to say $ug = 0$ for every g in $\operatorname{Hom}_\Lambda(C, A)$. Thus $ug(e) = 0$ for all such g. But we can arrange that $g(e)$ is any prescribed element of A. Consequently $u(a) = 0$ for all $a \in A$, and with this the theorem is proved.

Corollary. *Any non-zero free Λ-module is a projective generator for $\mathscr{C}_\Lambda$.*

We mention briefly the dual theory. Let $T : \mathscr{C}_\Lambda \to \mathscr{C}_\Delta$ be an additive *contravariant* functor, then when A and B are Λ-modules T induces a homomorphism
$$\operatorname{Hom}_\Lambda(A, B) \to \operatorname{Hom}_\Delta(T(B), T(A)) \qquad (4.3.2)$$
of abelian groups.

Definition. *The contravariant functor $T : \mathscr{C}_\Lambda \to \mathscr{C}_\Delta$ is said to be 'faithful' if (4.3.2) is a monomorphism for all choices of A and B.*

Let M be a Λ-module. Then $\operatorname{Hom}_\Lambda(-, M)$ is an additive contravariant functor from $\mathscr{C}_\Lambda$ to $\mathscr{C}_Z$. If this functor is faithful, then M is called a *cogenerator* of $\mathscr{C}_\Lambda$.

Exercise 3. *Let E be an injective Λ-module containing $\bigoplus_L \Lambda/L$, where L ranges over all the maximal left ideals of Λ. Show that E is an injective cogenerator of $\mathscr{C}_\Lambda$.*

4.4 Equivalent categories

It can happen that two categories of modules, say $\mathscr{C}_\Lambda$ and $\mathscr{C}_\Delta$, have almost identical properties even though the rings Λ and Δ differ in important respects. (For example one of the rings might be commutative and the other not.) However the resemblance between the two categories suggests that Λ and Δ will have similar homological properties. It is this idea that we shall investigate in the remainder of this chapter.

Suppose that $F:\mathscr{C}_\Lambda \to \mathscr{C}_\Delta$ is an additive covariant functor and that A, B are Λ-modules. Then, as we noted in section (4.3), F will induce a homomorphism $\quad \mathrm{Hom}_\Lambda\,(A, B) \to \mathrm{Hom}_\Delta\,(F(A), F(B)) \qquad (4.4.1)$ of abelian groups.

Definition. *The additive covariant functor $F:\mathscr{C}_\Lambda \to \mathscr{C}_\Delta$ is said to be 'fully faithful' if $(4.4.1)$ is an isomorphism for all choices of A and B.*

Exercise 4. *The additive covariant functor $F:\mathscr{C}_\Lambda \to \mathscr{C}_\Delta$ is fully faithful and $f:A \to B$ is a homomorphism in $\mathscr{C}_\Lambda$. Show that f is an isomorphism if and only if $F(f)$ is an isomorphism.*

We introduce a further definition.

Definition. *The covariant functor $F:\mathscr{C}_\Lambda \to \mathscr{C}_\Delta$ is called an 'equivalence' if it is additive and there exists an additive covariant functor $G:\mathscr{C}_\Delta \to \mathscr{C}_\Lambda$ such that GF is naturally equivalent† to the identity functor on $\mathscr{C}_\Lambda$ and FG is naturally equivalent to the identity functor on $\mathscr{C}_\Delta$.*

If there exists an equivalence from $\mathscr{C}_\Lambda$ to $\mathscr{C}_\Delta$, then we say that the categories $\mathscr{C}_\Lambda$ and $\mathscr{C}_\Delta$ are *equivalent*. Equivalence between categories is obviously reflexive and symmetric. An easy verification shows that it is also transitive. When F and G are as described in the definition, they are referred to as *inverse equivalences*.

Let $F:\mathscr{C}_\Lambda \to \mathscr{C}_\Delta$ and $G:\mathscr{C}_\Delta \to \mathscr{C}_\Lambda$ be covariant functors. For the moment we shall not make any assumptions about additivity. Denote by I_Λ respectively I_Δ the identity functor on $\mathscr{C}_\Lambda$ respectively $\mathscr{C}_\Delta$ and suppose that we are given natural equivalences $\eta:I_\Lambda \to GF$ and $\zeta:I_\Delta \to FG$. In this situation let $f:A \to B$ be a Λ-homomorphism. We then have a commutative diagram

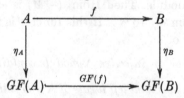

† See section (1.7) for the definition of naturally equivalent functors.

in which η_A and η_B are Λ-isomorphisms. It follows that the mapping which carries f into $GF(f)$ is a bijection of $\mathrm{Hom}_\Lambda\,(A,B)$ on to

$$\mathrm{Hom}_\Lambda\,(GF(A), GF(B)).$$

Consequently the mapping

$$\mathrm{Hom}_\Lambda\,(A,B) \to \mathrm{Hom}_\Lambda\,(F(A), F(B))$$

induced by F and the mapping

$$\mathrm{Hom}_\Delta\,(F(A), F(B)) \to \mathrm{Hom}_\Lambda\,(GF(A), GF(B))$$

induced by G are injective and surjective respectively.

Let K and M be Δ-modules and put $G(K) = A'$ and $G(M) = B'$. We then have a bijection between $\mathrm{Hom}_\Delta\,(K, M)$ and

$$\mathrm{Hom}_\Delta\,(F(A'), F(B'))$$

such that if g in $\mathrm{Hom}_\Delta\,(K, M)$ corresponds to ϕ in

$$\mathrm{Hom}_\Delta\,(F(A'), F(B')),$$

then the diagram

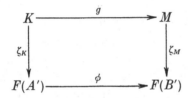

is commutative. Let g and ϕ correspond in this way. Then

$$
\begin{array}{ccc}
G(K) & \xrightarrow{\;\;G(g)\;\;} & G(M) \\
{\scriptstyle G(\zeta_K)}\downarrow & & \downarrow{\scriptstyle G(\zeta_M)} \\
GF(A') & \xrightarrow{\;\;G(\phi)\;\;} & GF(B')
\end{array}
$$

is also a commutative diagram and the vertical mappings are Λ-isomorphisms. Now the results of the last paragraph show that every Λ-homomorphism from $GF(A')$ to $GF(B')$ is of the form $G(\phi)$. Consequently every member of $\mathrm{Hom}_\Lambda\,(G(K), G(M))$ is of the form $G(g)$. However the relation between F and G is symmetrical and therefore a similar conclusion applies to the members of $\mathrm{Hom}_\Delta\,(F(A), F(B))$. Collecting together the information we have derived concerning F, we arrive at the following result: *the mapping*

$$\mathrm{Hom}_\Lambda\,(A,B) \to \mathrm{Hom}_\Delta\,(F(A), F(B))$$

induced by F is a bijection. In particular, if $F: \mathscr{C}_\Lambda \to \mathscr{C}_\Delta$ is an equivalence, then it is fully faithful.

Exercise 5. *Let* $F: \mathscr{C}_\Lambda \to \mathscr{C}_\Delta$ *be an additive covariant functor and let* $G: \mathscr{C}_\Delta \to \mathscr{C}_\Lambda$ *be a covariant functor in the reverse direction. Assume that* GF *resp. FG is naturally equivalent to the identity functor on* $\mathscr{C}_\Lambda$ *resp.* $\mathscr{C}_\Delta$. *Show that G is also additive.* (*Thus F and G are inverse equivalences.*)

We are now ready to give a useful characterization of equivalences.

Theorem 7. *Let* $F: \mathscr{C}_\Lambda \to \mathscr{C}_\Delta$ *be an additive covariant functor. Then in order that F should be an equivalence it is necessary and sufficient that the following two conditions be satisfied:*
- (a) *F is fully faithful;*
- (b) *to each module in* $\mathscr{C}_\Delta$ *there shall correspond an isomorphic module of the form F(A).*

Proof. We shall assume that (a) and (b) hold and deduce that F is an equivalence. The converse follows at once from what has already been established.

Let K belong to $\mathscr{C}_\Delta$ and, using (b), construct an isomorphism $\zeta_K: K \overset{\sim}{\to} F(A)$, where A is a suitable Λ-module. Define $G(K)$ by $G(K) = A$ so that $\zeta_K: K \to FG(K)$. Now suppose that $v: K \to K'$ in $\mathscr{C}_\Delta$. There then exists a unique Δ-homomorphism $\beta: FG(K) \to FG(K')$ with the property that $\beta \zeta_K = \zeta_{K'} v$. It follows, from (a), that there is a unique member of $\mathrm{Hom}_\Lambda (G(K), G(K'))$ which is transformed by F into β. This member will be denoted by $G(v)$. Accordingly

$$G(v): G(K) \to G(K')$$

is a Λ-homomorphism which is characterized by the property that the diagram

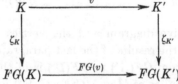

is commutative. Evidently $G(\text{identity}) = \text{identity}$. Also if $v': K' \to K''$ is another Δ-homomorphism, then $G(v'v) = G(v') G(v)$. Thus we have constructed a covariant functor $G: \mathscr{C}_\Delta \to \mathscr{C}_\Lambda$ which is such that the ζ_K provide a natural equivalence between the composite functor FG and the identity functor I_Δ.

Now suppose that A is a Λ-module. Then

$$\zeta_{F(A)}: F(A) \to FGF(A)$$

is a Δ-isomorphism. It follows, from (a), that there exists a unique homomorphism $\qquad \eta_A: A \to GF(A)$

in $\mathscr{C}_\Lambda$ such that $F(\eta_A) = \zeta_{F(A)}$. Further, by Exercise 4, η_A is an isomorphism. Let $u: A \to A'$ be a Λ-homomorphism and consider the diagram

$$
\begin{array}{ccc}
A & \xrightarrow{\quad u \quad} & A' \\
{\scriptstyle \eta_A}\big\downarrow & & \big\downarrow{\scriptstyle \eta_{A'}} \\
GF(A) & \xrightarrow{\quad GF(u) \quad} & GF(A')
\end{array}
\qquad (4.4.2)
$$

It is clear that, on applying F, the resulting diagram is commutative. It follows, again from (a), that (4.4.2) is itself commutative. Accordingly the η_A collectively constitute a natural equivalence between the identity functor I_Λ and GF. To complete the proof it suffices to show that G is additive. This, however, is a consequence of Exercise 5.

Exercise 6. *Let $F: \mathscr{C}_\Lambda \to \mathscr{C}_\Delta$ and $G: \mathscr{C}_\Delta \to \mathscr{C}_\Lambda$ be inverse equivalences. Show that if A is a Λ-module and K a Δ-module, then there exists an isomorphism*

$$\mathrm{Hom}_\Lambda\,(A, G(K)) \xrightarrow{\sim} \mathrm{Hom}_\Delta\,(F(A), K)$$

of abelian groups which is natural for homomorphisms $A' \to A$ and $K \to K'$ in $\mathscr{C}_\Lambda$ and $\mathscr{C}_\Delta$ respectively.

At this point the question arises as to the extent to which equivalent categories resemble one another. This will now be investigated.

Lemma 2. *Let $F: \mathscr{C}_\Lambda \to \mathscr{C}_\Delta$ be an equivalence and suppose that $\phi: A \to B$ and $\psi: B \to C$ are Λ-homomorphisms. Then*

(a) *the sequence $0 \to A \xrightarrow{\phi} B \xrightarrow{\psi} C$ is exact in $\mathscr{C}_\Lambda$ if and only if*

$$0 \to F(A) \xrightarrow{F(\phi)} F(B) \xrightarrow{F(\psi)} F(C)$$

is exact in $\mathscr{C}_\Delta$;

(b) *the sequence $A \xrightarrow{\phi} B \xrightarrow{\psi} C \to 0$ is exact in $\mathscr{C}_\Lambda$ if and only if*

$$F(A) \xrightarrow{F(\phi)} F(B) \xrightarrow{F(\psi)} F(C) \to 0$$

is exact in $\mathscr{C}_\Delta$.

Proof. First assume that the sequence

$$0 \to F(A) \xrightarrow{F(\phi)} F(B) \xrightarrow{F(\psi)} F(C)$$

is exact. Then $F(\psi\phi) = 0$ and therefore $\psi\phi = 0$ because, by Theorem 7, F is fully faithful. Now let $f: D \to B$ be a Λ-homomorphism such that $\psi f = 0$. Further let $g: D \to A$ be a typical member of $\mathrm{Hom}_\Lambda (D, A)$. In order to show† that $0 \to A \xrightarrow{\phi} B \xrightarrow{\psi} C$ is exact it will suffice to prove that there is a unique g such that $\phi g = f$. But $F(\psi) F(f) = 0$. Consequently there is a unique Δ-homomorphism $\gamma: F(D) \to F(A)$ such that $F(f) = F(\phi)\gamma$. But, by Theorem 7, $\phi g = f$ if and only if

$$F(f) = F(\phi) F(g)$$

that is if and only if $F(g) = \gamma$. A further appeal to Theorem 7 shows that there is a unique g with this property.

Next suppose that $0 \to A \xrightarrow{\phi} B \xrightarrow{\psi} C$ is exact and let $G: \mathscr{C}_\Delta \to \mathscr{C}_\Lambda$ be an equivalence inverse to F. Then $0 \to GF(A) \to GF(B) \to GF(C)$ is exact. The results of the last paragraph applied to the equivalence G allow us to conclude that

$$0 \to F(A) \to F(B) \to F(C)$$

is exact. This disposes of (a) and (b) may be dealt with similarly.

Theorem 8. *If $F: \mathscr{C}_\Lambda \to \mathscr{C}_\Delta$ is an equivalence, then the functor F is exact.*

Proof. This is an immediate consequence of Lemma 2. However, as the next theorem shows, we can say a great deal more.

Theorem 9. *Let $F: \mathscr{C}_\Lambda \to \mathscr{C}_\Delta$ be an equivalence and let $\phi: A \to B$ and $\psi: B \to C$ be Λ-homomorphisms. Then $A \xrightarrow{\phi} B \xrightarrow{\psi} C$ is exact in $\mathscr{C}_\Lambda$ if and only if $F(A) \xrightarrow{F(\phi)} F(B) \xrightarrow{F(\psi)} F(C)$ is exact in $\mathscr{C}_\Delta$.*

Proof. We shall assume that $F(A) \to F(B) \to F(C)$ is exact and deduce that $A \to B \to C$ is exact. The converse follows by virtue of Theorem 8.

Let $G: \mathscr{C}_\Delta \to \mathscr{C}_\Lambda$ be an equivalence inverse to F. Then, by Theorem 8, G is exact and therefore $GF(A) \to GF(B) \to GF(C)$ is an exact sequence. However this obviously implies that $A \to B \to C$ is exact as required.

Exercise 7. *Let $F: \mathscr{C}_\Lambda \to \mathscr{C}_\Delta$ be an equivalence and let K be a Λ-module. Show that K is a generator of $\mathscr{C}_\Lambda$ if and only if $F(K)$ is a generator of $\mathscr{C}_\Delta$.*

† See Chapter 1, Supplementary Exercise C.

Theorem 10. *Let* $F:\mathscr{C}_\Lambda \to \mathscr{C}_\Delta$ *be an equivalence and let* A,B *be* Λ-*modules. Then*

(a) A *is* Λ-*projective if and only if* $F(A)$ *is* Δ-*projective;*

(b) B *is* Λ-*injective if and only if* $F(B)$ *is* Δ-*injective.*

Proof. Let $G:\mathscr{C}_\Delta \to \mathscr{C}_\Lambda$ be an equivalence inverse to F. We shall make use of the natural isomorphism

$$\mathrm{Hom}_\Lambda\,(A, G(K)) \approx \mathrm{Hom}_\Delta\,(F(A), K) \qquad (4.4.3)$$

introduced in Exercise 6.

Suppose that A is Λ-projective and let $0 \to K' \to K \to K'' \to 0$ be an exact sequence in $\mathscr{C}_\Delta$. Since G is exact (Theorem 8), the induced sequence $0 \to G(K') \to G(K) \to G(K'') \to 0$ is exact in $\mathscr{C}_\Lambda$ and therefore

$$0 \to \mathrm{Hom}_\Lambda\,(A, G(K')) \to \mathrm{Hom}_\Lambda\,(A, G(K)) \to \mathrm{Hom}_\Lambda\,(A, G(K'')) \to 0$$

is also exact because A is Λ-projective. It follows, from (4.4.3), that the sequence

$$0 \to \mathrm{Hom}_\Delta\,(F(A), K') \to \mathrm{Hom}_\Delta\,(F(A), K) \to \mathrm{Hom}_\Delta\,(F(A), K'') \to 0$$

is exact as well. This proves that $F(A)$ is Δ-projective.

Next suppose that $F(A)$ is Δ-projective. Then the reasoning just given shows that $GF(A)$ is Λ-projective. But A and $GF(A)$ are Λ-isomorphic. Consequently A is Λ-projective and now (a) is proved. A simple adaptation of the argument yields a proof of (b).

Theorem 11. *Let* $F:\mathscr{C}_\Lambda \to \mathscr{C}_\Delta$ *be an equivalence and let* A,B *be* Λ-*modules. Then*

(1) *the projective dimensions of* A *and* $F(A)$ *are equal;*

(2) *the injective dimensions of* B *and* $F(B)$ *are equal.*

Proof. We use our customary notation for projective dimension and begin by showing that $\mathrm{Pd}_\Delta\,(F(A)) \leqslant \mathrm{Pd}_\Lambda\,(A)$. For this we may suppose that $\mathrm{Pd}_\Lambda\,(A) = n$, where $0 \leqslant n < \infty$. In these circumstances there exists an exact sequence

$$0 \to P_n \to P_{n-1} \to \dots \to P_1 \to P_0 \to A \to 0,$$

where each P_ν is Λ-projective. But, by Theorem 8, F is exact. Consequently we have, in $\mathscr{C}_\Delta$, an exact sequence

$$0 \to F(P_n) \to F(P_{n-1}) \to \dots \to F(P_1) \to F(P_0) \to F(A) \to 0$$

and, by Theorem 10, $F(P_\nu)$ is Δ-projective. This shows that

$$\mathrm{Pd}_\Delta\,(F(A)) \leqslant n = \mathrm{Pd}_\Lambda\,(A)$$

and establishes the desired inequality.

Let $G: \mathscr{C}_\Delta \to \mathscr{C}_\Lambda$ be an equivalence inverse to F. Then $GF(A)$ and A are Λ-isomorphic modules and our previous considerations show that $\mathrm{Pd}_\Lambda(GF(A)) \leqslant \mathrm{Pd}_\Delta(F(A))$. Accordingly $\mathrm{Pd}_\Lambda(A) \leqslant \mathrm{Pd}_\Delta(F(A))$ and with this (1) is proved. Similar considerations yield (2).

From (Chapter 3, Theorem 17) we know that

$$\sup_{A \in \mathscr{C}_\Lambda} \mathrm{Pd}_\Lambda(A) = \sup_{B \in \mathscr{C}_\Lambda} \mathrm{Id}_\Lambda(B).$$

Put
$$GD(\mathscr{C}_\Lambda) = \sup_{A \in \mathscr{C}_\Lambda} \mathrm{Pd}_\Lambda(A) = \sup_{B \in \mathscr{C}_\Lambda} \mathrm{Id}_\Lambda(B) \qquad (4.4.4)$$

so that, with the notation of section (3.7),

$$GD(\mathscr{C}_\Lambda^L) = l.GD(\Lambda) \qquad (4.4.5)$$

and
$$GD(\mathscr{C}_\Lambda^R) = r.GD(\Lambda). \qquad (4.4.6)$$

Theorem 12. *Let $F: \mathscr{C}_\Lambda \to \mathscr{C}_\Delta$ be an equivalence. Then*

$$GD(\mathscr{C}_\Lambda) = GD(\mathscr{C}_\Delta).$$

Proof. Every Δ-module K is isomorphic to a module of the form $F(A)$ and, for such an $A, \mathrm{Pd}_\Delta(K) = \mathrm{Pd}_\Delta(F(A))$. Hence, by Theorem 11,

$$GD(\mathscr{C}_\Delta) = \sup_{A \in \mathscr{C}_\Lambda} \mathrm{Pd}_\Delta(F(A)) = \sup_{A \in \mathscr{C}_\Lambda} \mathrm{Pd}_\Lambda(A) = GD(\mathscr{C}_\Lambda).$$

We shall now digress a little in order to prepare the way for an application of these results. Let B be a Λ-module. Then $\mathrm{Hom}_\Lambda(B, B)$ is certainly an abelian group. Suppose that f and g belong to

$$\mathrm{Hom}_\Lambda(B, B).$$

Then each of them is a Λ-homomorphism of B into itself. Accordingly gf, that is the Λ-homomorphism which maps the element $b \in B$ into $g(f(b))$, also belongs to $\mathrm{Hom}_\Lambda(B, B)$. Thus the elements of $\mathrm{Hom}_\Lambda(B, B)$ may be multiplied as well as added and indeed they form a ring. This ring, which is called the *endomorphism* ring of B, will be denoted by $\mathrm{End}_\Lambda(B)$. It has i_B as its identity element and the zero element is the null homomorphism of B into itself. We recall that an element of a ring is called a *unit* if it has a two-sided inverse. The units of $\mathrm{End}_\Lambda(B)$ are the automorphisms of B.

Exercise 8.* *Let E be a non-zero, injective Λ-module. Show that the following statements are equivalent:*

(a) *E has no non-trivial direct summands;*

(b) *E is the injective envelope of each of its non-zero submodules;*

(c) *every pair of non-zero submodules of E has a non-zero intersection;*
(d) *the non-units of* $\mathrm{End}_\Lambda(E)$ *form a two-sided ideal.*

For the reader's interest we mention that a non-zero injective module E having the four equivalent properties listed in Exercise 8 is called an *indecomposable injective*. We return to the main discussion.

In order to be more explicit let us work in the category $\mathscr{C}_\Lambda^R$. Then B will be a right Λ-module. Put $\Omega = \mathrm{End}_\Lambda(B)$ and, if $b \in B$ and $\phi \in \Omega$, write

$$\phi b = \phi(b).$$

It is easy to check that this turns B into a left Ω-module, i.e. B belongs to $\mathscr{C}_\Omega^L$. In fact if $\phi \in \Omega$ and $\lambda \in \Lambda$, then

$$\phi(b\lambda) = (\phi(b))\lambda = (\phi b)\lambda.$$

In other terms, B is a (Λ, Ω)-bimodule with Λ operating on the right and Ω on the left. Now the Hom functor is contravariant in its first variable. It follows that $\mathrm{Hom}_\Lambda(B, -)$ is an additive covariant functor from $\mathscr{C}_\Lambda^R$ to $\mathscr{C}_\Omega^R$. Note that if $f \in \mathrm{Hom}_\Lambda(B, A)$ and $\phi \in \Omega$, then the product of f and ϕ with respect to the structure that $\mathrm{Hom}_\Lambda(B, A)$ possesses as a right Ω-module, is simply the composition $f\phi$ of the homomorphisms $\phi : B \to B$ and $f : B \to A$. *In particular, taking $A = B$,* $\mathrm{Hom}_\Lambda(B, B)$ *considered as a right Ω-module is just the ring Ω regarded as a right module with respect to itself in the usual way.*

Theorem 13. *Let B be a right Λ-module and put $\Omega = \mathrm{End}_\Lambda(B)$. Further let $\mathrm{Hom}_\Lambda(B, -) = F$ (say) be considered as a covariant functor from $\mathscr{C}_\Lambda^R$ to $\mathscr{C}_\Omega^R$. Then the mapping*

$$\Omega = \mathrm{Hom}_\Lambda(B, B) \to \mathrm{Hom}_\Omega(F(B), F(B)) = \mathrm{End}_\Omega(\Omega) \quad (4.4.7)$$

induced by F is a ring-isomorphism.

Proof. Let ϕ, ϕ_1 and ϕ_2 belong to $\mathrm{Hom}_\Lambda(B, B) = \Omega$. Then

$$F(\phi_1 + \phi_2) = F(\phi_1) + F(\phi_2),$$

because F is additive, and $F(i_B) = i_\Omega$ because F is a functor. Next $F(\phi) = \mathrm{Hom}_\Lambda(B, \phi)$ and therefore, if $\omega \in \Omega$,

$$(F(\phi))(\omega) = \phi\omega. \quad (4.4.8)$$

Accordingly

$$(F(\phi_1\phi_2))(\omega) = \phi_1\phi_2\omega = \phi_1((F(\phi_2))(\omega))$$

whence $F(\phi_1\phi_2) = F(\phi_1)F(\phi_2)$. This shows that (4.4.7) is a ring-homomorphism and, by (4.4.8), it is clearly an injective mapping.

Finally let $\alpha \in \mathrm{Hom}_\Omega(\Omega, \Omega)$ it being understood that α is in $\mathscr{C}_\Omega^R$. Put $\psi = \alpha(1_\Omega)$. Then $\psi \in \mathrm{Hom}_\Lambda(B, B)$ and for $\omega \in \Omega$ we have

$$(F(\psi))(\omega) = \psi\omega = \alpha(1_\Omega)\omega = \alpha(\omega)$$

whence $F(\psi) = \alpha$. Thus our ring-homomorphism is also a surjective mapping and now the proof is complete.

For the next theorem we assume that B is a right Λ-module and that $\{A_i\}_{i \in I}$ is a family of right Λ-modules. The direct sum of the A_i is denoted by A and $\pi_i : A \to A_i$ is used to indicate the canonical projection of A on to the summand A_i.

Theorem 14. *Let the situation be as described above. If the Λ-module B is finitely generated, then there is an isomorphism*

$$\mathrm{Hom}_\Lambda(B, A) \approx \bigoplus_{i \in I} \mathrm{Hom}_\Lambda(B, A_i)$$

(of abelian groups) such that f in $\mathrm{Hom}_\Lambda(B, A)$ corresponds to $\{\pi_i f\}_{i \in I}$ in $\bigoplus_{i \in I} \mathrm{Hom}_\Lambda(B, A_i)$. If $\Omega = \mathrm{End}_\Lambda(B)$, then the same mapping is also an isomorphism of right Ω-modules.

Proof. The fact that B is finitely generated ensures that $\pi_i f$ is a null homomorphism for almost all i and hence that $\{\pi_i f\}_{i \in I}$ belongs to $\bigoplus_{i \in I} \mathrm{Hom}_\Lambda(B, A_i)$. The mapping which takes f into $\{\pi_i f\}_{i \in I}$ is a homomorphism of abelian groups which is obviously a monomorphism. Let $\{g_i\}_{i \in I}$ belong to $\bigoplus_{i \in I} \mathrm{Hom}_\Lambda(B, A_i)$ and, for $b \in B$, put $g(b) = \{g_i(b)\}_{i \in I}$. Then $g \in \mathrm{Hom}_\Lambda(B, A)$ and it is mapped into $\{g_i\}_{i \in I}$. This proves that we are dealing with an isomorphism of abelian groups.

Finally assume that $f \in \mathrm{Hom}_\Lambda(B, A)$ and that $\phi \in \Omega$. As already observed, the product of f and ϕ when $\mathrm{Hom}_\Lambda(B, A)$ is regarded as a right Ω-module is the composite homomorphism $f\phi$. If the isomorphism is applied to this product, then we obtain $\{\pi_i f\phi\}_{i \in I}$ which is none other than the product of $\{\pi_i f\}_{i \in I}$ and ϕ. Accordingly the isomorphism is not only an isomorphism of abelian groups, but also an isomorphism of Ω-modules.

4.5 Matrix rings

In this section we shall be working primarily in $\mathscr{C}_\Lambda^R$. If A is a Λ-module and I is an arbitrary set, then as in previous similar situations $\bigoplus_I A$ will denote a direct sum in which all the summands are equal to A

and there is one of them for each member of I. Likewise $\prod_I A$ will denote the direct product in which every factor is A and there is one factor for each element of the set I.

Lemma 3. *Let B be a finitely generated, right Λ-module and put $\Omega = \mathrm{End}_\Lambda(B)$. Suppose that $C = \bigoplus_I B$ and $D = \bigoplus_J B$. Denote by F the additive functor from $\mathscr{C}_\Lambda^R$ to $\mathscr{C}_\Omega^R$ given by $F = \mathrm{Hom}_\Lambda(B, -)$. Then the homomorphism*
$$\mathrm{Hom}_\Lambda(C, D) \to \mathrm{Hom}_\Omega(F(C), F(D))$$
(of abelian groups) induced by F is an isomorphism.

Proof. We shall show that the lemma can be reduced to the case where $C = B$ and $D = B$. In this situation the desired result follows by Theorem 13.

Let $\sigma_i: B \to C$ resp. $\pi_i: C \to B$ be the ith injection resp. projection homomorphism. By Theorem 14, there exists an Ω-isomorphism
$$\mathrm{Hom}_\Lambda(B, C) \xrightarrow{\sim} \bigoplus_I \mathrm{Hom}_\Lambda(B, B),$$
that is an Ω-isomorphism
$$F(C) \xrightarrow{\sim} \bigoplus_I F(B),$$
in which f in $\mathrm{Hom}_\Lambda(B, C)$ corresponds to $\{\pi_i f\}_{i \in I} = \{(F(\pi_i))(f)\}_{i \in I}$ in $\bigoplus_I \mathrm{Hom}_\Lambda(B, B)$. It follows that, for each $i \in I$, the diagram

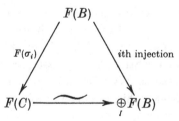

is commutative. Using this together with (Chapter 2, Theorem 2) we obtain isomorphisms (of abelian groups)
$$\mathrm{Hom}_\Omega(F(C), F(D)) \approx \mathrm{Hom}_\Omega\left(\bigoplus_I F(B), F(D)\right) \approx \prod_I \mathrm{Hom}_\Omega(F(B), F(D)).$$
To see how these operate, suppose that ϕ in $\mathrm{Hom}_\Omega(F(C), F(D))$ corresponds to ψ in $\mathrm{Hom}_\Omega\left(\bigoplus_I F(B), F(D)\right)$ and to
$$\{\phi_i\}_{i \in I} \quad \text{in} \quad \prod_I \mathrm{Hom}_\Omega(F(B), F(D)).$$

Then we have a commutative diagram

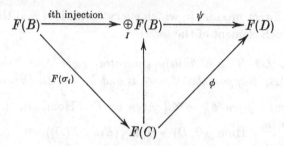

and, by (Chapter 2, Theorem 2), ϕ_i is the result of combining the two mappings in the upper row. Accordingly ϕ in $\mathrm{Hom}_\Omega\,(F(C), F(D))$ corresponds to $\{\phi F(\sigma_i)\}_{i \in I}$ in $\prod\limits_I \mathrm{Hom}_\Omega\,(F(B), F(D))$. Again

$$\mathrm{Hom}_\Lambda\,(C, D) = \mathrm{Hom}_\Lambda\,(\bigoplus_I B, C) \approx \prod_I \mathrm{Hom}_\Lambda\,(B, D),$$

where g in $\mathrm{Hom}_\Lambda\,(C, D)$ corresponds to $\{g\sigma_i\}_{i \in I}$ in $\prod\limits_I \mathrm{Hom}_\Lambda\,(B, D)$.

Consider the diagram

$$
\begin{array}{ccc}
\mathrm{Hom}_\Lambda(C, D) & \xrightarrow{\;\sim\;} & \prod\limits_I \mathrm{Hom}_\Lambda(B, D) \\
\downarrow & & \downarrow \\
\mathrm{Hom}_\Omega(F(C), F(D)) & \xrightarrow{\;\sim\;} & \prod\limits_I \mathrm{Hom}_\Omega(F(B), F(D))
\end{array}
$$

Here the horizontal mappings are the isomorphisms already encountered. The left vertical mapping is induced by F and the right vertical mapping is also induced by F but this time on a term by term basis. The observations of the last paragraph show that the diagram is commutative. We wish to show that the vertical mapping on the left is an isomorphism. This will follow if we can prove that the homomorphism

$$\mathrm{Hom}_\Lambda\,(B, D) \to \mathrm{Hom}_\Omega\,(F(B), F(D)),$$

induced by F, is an isomorphism. Accordingly we may make an entirely fresh start and suppose from here on that $C = B$.

Let $p_j \colon D \to B$ be the jth projection. Since B is finitely generated we have, by Theorem 14, an Ω-isomorphism

$$\mathrm{Hom}_\Lambda\,(B, D) \xrightarrow{\;\sim\;} \bigoplus_J \mathrm{Hom}_\Lambda\,(B, B) \qquad (4.5.1)$$

in which f in $\mathrm{Hom}_\Lambda (B, D)$ corresponds to $\{p_j f\}_{j \in J} = \{(F(p_j))\,(f)\}_{j \in J}$ in $\underset{J}{\oplus}\, \mathrm{Hom}_\Lambda (B, B)$. It follows that, for each j in J, the diagram

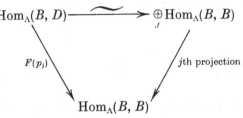

is commutative, that is to say

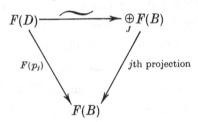

is a commutative diagram in $\mathscr{C}_\Omega^R$. Using the horizontal mapping we arrive at isomorphisms

$$\mathrm{Hom}_\Omega\,(F(B), F(D)) \approx \mathrm{Hom}_\Omega\,(F(B), \underset{J}{\oplus}\, F(B)) \approx \underset{J}{\oplus}\, \mathrm{Hom}_\Omega\,(F(B), F(B))$$

of abelian groups. Note that the second isomorphism follows from Theorem 14 as soon as it is observed that $F(B)$ is a finitely generated Ω-module. (In fact $F(B)$ is Ω itself considered as a right Ω-module.) To see how the isomorphisms work, let ϕ in $\mathrm{Hom}_\Omega\,(F(B), F(D))$ correspond to ψ in $\mathrm{Hom}_\Omega\,(F(B), \underset{J}{\oplus}\, F(B))$ and to $\{\phi_j\}_{j \in J}$ in

$$\underset{J}{\oplus}\, \mathrm{Hom}_\Omega\,(F(B), F(B)).$$

Then

$$F(B) \xrightarrow{\ \psi\ } \underset{J}{\oplus}\, F(B) \xrightarrow{\ \text{jth projection}\ } F(B)$$

is a commutative diagram and the result of composing the two homomorphisms in the upper row is ϕ_j. Accordingly the member of $\underset{J}{\oplus}\, \mathrm{Hom}_\Omega\,(F(B), F(B))$ that corresponds to ϕ is $\{F(p_j)\,\phi\}_{j \in J}$.

Consider the diagram

$$\begin{array}{ccc} \mathrm{Hom}_\Lambda(B, D) & \xrightarrow{\;\approx\;} & \underset{J}{\oplus}\,\mathrm{Hom}_\Lambda(B, B) \\ \downarrow & & \downarrow \\ \mathrm{Hom}_\Omega(F(B), F(D)) & \xrightarrow{\;\approx\;} & \underset{J}{\oplus}\,\mathrm{Hom}_\Omega(F(B), F(B)) \end{array}$$

where the horizontal mappings are the isomorphisms previously investigated and both the vertical mappings are induced by F. The observations of the last paragraph show that the diagram is commutative. By Theorem 13, the homomorphism

$$\mathrm{Hom}_\Lambda (B, B) \to \mathrm{Hom}_\Omega (F(B), F(B)),$$

which F induces, is an isomorphism. Accordingly

$$\mathrm{Hom}_\Lambda (B, D) \to \mathrm{Hom}_\Omega (F(B), F(D))$$

is also an isomorphism and with this the proof of the lemma is complete.

We come now to the main result of this section.

Theorem 15. *Let B be a finitely generated, projective generator of $\mathscr{C}_\Lambda^R$. Put $F = \mathrm{Hom}_\Lambda (B, -)$ and $\Omega = \mathrm{End}_\Lambda (B)$. Then $F : \mathscr{C}_\Lambda^R \to \mathscr{C}_\Omega^R$ is an equivalence.*

Proof. Let I be any set. By Theorem 14,

$$F(\underset{I}{\oplus} B) = \mathrm{Hom}_\Lambda (B, \underset{I}{\oplus} B) \approx \underset{I}{\oplus} \mathrm{Hom}_\Lambda (B, B) = \underset{I}{\oplus} \Omega$$

and the isomorphism in the middle is an isomorphism of Ω-modules. Accordingly $F(\underset{I}{\oplus} B)$ is a free right Ω-module and every free right Ω-module is isomorphic to a module of this form. The functor F is additive. We shall show that it is an equivalence by making use of the criterion provided by Theorem 7.

Let M be a right Ω-module. It is possible to construct an exact sequence

$$F(\underset{J}{\oplus} B) \xrightarrow{\phi} F(\underset{I}{\oplus} B) \to M \to 0$$

of Ω-modules. By Lemma 3, there exists a Λ-homomorphism

$$f : \underset{J}{\oplus} B \to \underset{I}{\oplus} B$$

such that $F(f) = \phi$. Using f, we construct an exact sequence

$$\underset{J}{\oplus} B \xrightarrow{f} \underset{I}{\oplus} B \to A \to 0$$

in $\mathscr{C}_\Lambda$ and we apply the functor F to this sequence bearing in mind that F is exact because B is projective. In this way we arrive at the exact sequence

$$F(\underset{J}{\oplus} B) \overset{\phi}{\to} F(\underset{I}{\oplus} B) \to F(A) \to 0$$

whence $F(A)$ and M are isomorphic Ω-modules.

Let C and D be Λ-modules. In view of Theorem 7 the proof will be complete if we show that the homomorphism

$$\text{Hom}_\Lambda (C, D) \to \text{Hom}_\Omega (F(C), F(D)) \qquad (4.5.2)$$

induced by F is an isomorphism. But B is a generator of $\mathscr{C}_\Lambda^R$ and $\text{Hom}_\Omega (F(C), F(D))$ is a subgroup of $\text{Hom}_Z (F(C), F(D))$. This shows that (4.5.2) is a monomorphism, i.e. it shows that F, considered as a functor from $\mathscr{C}_\Lambda^R$ to $\mathscr{C}_\Omega^R$ (rather than from $\mathscr{C}_\Lambda^R$ to $\mathscr{C}_Z$), is faithful. Let $\psi : F(C) \to F(D)$ be an Ω-homomorphism. The proof will be complete if we show that there is a member of $\text{Hom}_\Lambda (C, D)$ which is mapped by F into ψ.

By Theorem 5 we can construct, in $\mathscr{C}_\Lambda^R$, exact sequences

$$\underset{I_2}{\oplus} B \overset{u_2}{\to} \underset{I_1}{\oplus} B \overset{u_1}{\to} C \to 0$$

and

$$\underset{J_2}{\oplus} B \overset{v_2}{\to} \underset{J_1}{\oplus} B \overset{v_1}{\to} D \to 0.$$

Bearing in mind that F is exact and that, for any set I, $F(\underset{I}{\oplus} B)$ is Ω-free, we observe that we can produce,† in $\mathscr{C}_\Omega^R$, a commutative diagram

with exact rows. But, by Lemma 3, there exist Λ-homomorphisms

$$\omega_1 : \underset{I_1}{\oplus} B \to \underset{J_1}{\oplus} B$$

and

$$\omega_2 : \underset{I_2}{\oplus} B \to \underset{J_2}{\oplus} B$$

such that $F(\omega_1) = \xi$ and $F(\omega_2) = \eta$. Accordingly $F(\omega_1 u_2) = F(v_2 \omega_2)$ and therefore $\omega_1 u_2 = v_2 \omega_2$ because, as we have already seen, F is

† See Chapter 2, Exercise 5.

faithful. It follows that there exists a Λ-homomorphism $g: C \to D$ such that the diagram

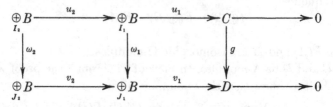

is commutative. Finally

$$F(g)\,F(u_1) = F(v_1)\,F(\omega_1) = F(v_1)\,\xi = \psi F(u_1)$$

whence $F(g) = \psi$ because $F(u_1)$ is an epimorphism. This ends the proof.

Let $n \geqslant 1$ be an integer and denote by $M_n(\Lambda)$ the ring formed by all square matrices of order n with elements in Λ. Further let B be a free right Λ-module with a base $e_1, e_2, \ldots, e_n$ of n elements. If now

$$f \in \mathrm{Hom}_\Lambda\,(B, B),$$

then

$$f(e_k) = \sum_{j=1}^{n} e_j a_{jk},$$

where the a_{jk} belong to Λ. Let us associate with f the element

$$\left\| \begin{matrix} a_{11} & a_{12} & \cdots & a_{1n} \\ a_{21} & a_{22} & \cdots & a_{2n} \\ \cdot & \cdot & & \cdot \\ \cdot & \cdot & & \cdot \\ \cdot & \cdot & & \cdot \\ a_{n1} & a_{n2} & \cdots & a_{nn} \end{matrix} \right\|$$

of $M_n(\Lambda)$. This association gives us a ring-isomorphism

$$\mathrm{End}_\Lambda\,(B) \approx M_n\,(\Lambda).$$

Since B is a finitely generated, projective generator for $\mathscr{C}_\Lambda^R$, the next theorem is an immediate consequence of Theorem 15.

Theorem 16. *Let $n \geqslant 1$ be an integer. Then the categories $\mathscr{C}_\Lambda^R$ and $\mathscr{C}_{M_n(\Lambda)}^R$ are equivalent and therefore $r.\mathrm{GD}\,(\Lambda) = r.\mathrm{GD}\,(M_n(\Lambda))$.*

We add a few remarks in conclusion. Note that if Λ is commutative and non-trivial, then $M_n(\Lambda)$ is non-commutative if $n \geqslant 2$. Nevertheless Λ and $M_n(\Lambda)$ give rise to equivalent categories. Again we have been working in this section with right Λ-modules instead of the more usual left modules. This is because it gives us a slight notational advantage. It is easy (but tiresome) to check that similar arguments hold for left Λ-modules. An alternative procedure is outlined below.

A bijection $\phi:\Lambda\to\Delta$ is called an *anti-isomorphism* if

$$\phi(\lambda_1+\lambda_2) = \phi(\lambda_1)+\phi(\lambda_2) \quad \text{and} \quad \phi(\lambda_1\lambda_2) = \phi(\lambda_2)\phi(\lambda_1)$$

for all λ_1, λ_2 in Λ. (Note that in these circumstances ϕ will necessarily map the identity element of Λ into the identity element of Δ.) If an anti-isomorphism exists between Λ and Δ, then the rings are said to be *anti-isomorphic*. This relation is symmetric.

Denote by Λ^* the ring which has the same elements as Λ and the same addition, but in which multiplication is defined by reversing the order of the two factors. We call Λ^* the *opposite* of Λ. Note that the identity mapping provides an anti-isomorphism between a ring and its opposite.

Exercise 9. *Show that if Λ and Δ are anti-isomorphic, then $\mathscr{C}_\Lambda^L$ and $\mathscr{C}_\Delta^R$ are equivalent. Deduce that $\mathscr{C}_\Lambda^L$ and $\mathscr{C}_{M_n(\Lambda)}^L$ are equivalent, where n is an arbitrary positive integer.*

Theorem 16 provides us with information concerning the global dimensions of a ring of the form $M_n(\Lambda)$. Now in the theory which deals with the structure of rings, direct sums of matrix rings of this type play an important role. This prompts a general question concerning the global dimensions of a direct sum of rings. The question and its answer are covered by the next exercise.

Exercise 10.* *The ring Λ is the direct sum of the rings $\Lambda_1, \Lambda_2, ..., \Lambda_s$. Show that*
$$r.\operatorname{GD}(\Lambda) = \max_{1\leqslant\mu\leqslant s} r.\operatorname{GD}(\Lambda_\mu).$$

Finally we mention, in passing, that the methods used in sections (4.4) and (4.5) can be used for deciding† whether a given abelian category is equivalent to a category $\mathscr{C}_\Lambda^R$ for some ring Λ.

Solutions to the Exercises on Chapter 4

Exercise 1. *Show that the polynomial functor is exact. Show also that if the left Λ-module A is free, then $A[x]$ is a free $\Lambda[x]$-module. Deduce that a left Λ-module B is Λ-projective if and only if $B[x]$ is $\Lambda[x]$-projective.*

† See (26), Theorem 1.19.

Solution. That the polynomial functor is exact is clear. Let F be a free Λ-module with a base $\{e_i\}_{i \in I}$. Then each e_i determines a corresponding constant polynomial $\bar{e}_i$ in $F[x]$. An easy verification shows that $\{\bar{e}_i\}_{i \in I}$ is a base for $F[x]$ over $\Lambda[x]$. In particular, $F[x]$ is a free $\Lambda[x]$-module.

Suppose that A is a projective Λ-module. Then A is a direct summand of a free Λ-module F. Applying the polynomial functor to the inclusion mapping $A \to F$, we see that $A[x]$ is a direct summand of the $\Lambda[x]$-module $F[x]$. However we know, from the last paragraph, that $F[x]$ is $\Lambda[x]$-free. Consequently $A[x]$ is $\Lambda[x]$-projective.

We obtain a ring-homomorphism $\Lambda \to \Lambda[x]$ by mapping each element of Λ into the corresponding constant polynomial. This enables us to regard each $\Lambda[x]$-module as a Λ-module. Considered as a Λ-module, $\Lambda[x]$ is free with the powers of x forming a base. It follows that each free $\Lambda[x]$-module is free when regarded as a Λ-module.

Now assume that A is a Λ-module and that $A[x]$ is $\Lambda[x]$-projective. Then $A[x]$ is a direct summand of a free $\Lambda[x]$-module Φ. If we regard $A[x]$ and Φ as Λ-modules, then $A[x]$ continues to be a direct summand of Φ and, by the last paragraph, Φ is Λ-free. Thus $A[x]$ is Λ-projective. But, as a Λ-module, $A[x]$ is a direct sum of copies of A. It follows, from (Chapter 2, Theorem 4), that A itself is Λ-projective.

Exercise 2. *Let Λ be a non-trivial ring and let $x_1, x_2, \ldots, x_s$ be indeterminates. Put $\Lambda[x] = \Lambda[x_1, x_2, \ldots, x_s]$. Show that*

$$l.\mathrm{Pd}_{\Lambda[x]} \left(\Lambda[x] / (x_1 \Lambda[x] + x_2 \Lambda[x] + \ldots + x_s \Lambda[x]) \right) = s.$$

Solution. Each x_i belongs to the centre of $\Lambda[x]$ and an easy verification shows that $x_1, x_2, \ldots, x_s$ is a $\Lambda[x]$-sequence in the sense of section (3.8). Further, because Λ is non-trivial,

$$\Lambda[x] \neq x_1 \Lambda[x] + x_2 \Lambda[x] + \ldots + x_s \Lambda[x].$$

The desired result therefore follows from (Chapter 3, Theorem 22).

Consider the ring-homomorphism $\Lambda[x] \to \Lambda$ in which each polynomial is mapped on to its constant term. This is surjective and $x_1 \Lambda[x] + x_2 \Lambda[x] + \ldots + x_s \Lambda[x]$ is its kernel. Hence if we use the ring-homomorphism to enable us to regard Λ as a $\Lambda[x]$-module, then Λ and $\Lambda[x] / (x_1 \Lambda[x] + x_2 \Lambda[x] + \ldots + x_s \Lambda[x])$ are isomorphic. Consequently our result may be stated as $l.\mathrm{Pd}_{\Lambda[x]} (\Lambda) = s$.

Exercise 3. *Let E be an injective module containing $\bigoplus_L \Lambda/L$, where L ranges over all the maximal left ideals of Λ. Show that E is an injective cogenerator of $\mathscr{C}_\Lambda^L$.*

Solution. Let A and B be left Λ-modules. We are required to show that the Z-homomorphism

$$\operatorname{Hom}_\Lambda(A,B) \to \operatorname{Hom}_Z(\operatorname{Hom}_\Lambda(B,E), \operatorname{Hom}_\Lambda(A,E))$$

induced by $\operatorname{Hom}_\Lambda(-,E)$ is a monomorphism.

Suppose that $f\colon A \to B$ is a non-zero Λ-homomorphism. Then there exists $a \in A$ such that $b = f(a)$ is non-zero. Let C be the submodule of B generated by b. Then the function $g\colon\Lambda \to C$ defined by $g(\lambda) = \lambda b$ is an epimorphism and hence C is isomorphic to $\Lambda/\operatorname{Ker} g$. But $\operatorname{Ker} g$ is a proper left ideal of Λ and therefore it is contained in a maximal left ideal, L say. The identity mapping of Λ induces an epimorphism $\Lambda/\operatorname{Ker} g \to \Lambda/L$. Since $\Lambda/L \subseteq E$ and C is isomorphic to $\Lambda/\operatorname{Ker} g$, we now have a Λ-homomorphism $h\colon C \to E$ for which $h(b) \neq 0$. Further $C \subseteq B$ and E is injective. Consequently there exists a Λ-homomorphism $h'\colon B \to E$ such that $h'j = h$, where $j\colon C \to B$ is an inclusion mapping. Thus $h'(b) \neq 0$ and therefore $h'f \neq 0$. Since $h'f$ is just $\operatorname{Hom}_\Lambda(f,E)$ applied to h', this shows that $\operatorname{Hom}_\Lambda(f,E) \neq 0$. The desired result follows.

Exercise 4. *The additive covariant functor $F\colon \mathscr{C}_\Lambda \to \mathscr{C}_\Delta$ is fully faithful and $f\colon A \to B$ is a Λ-homomorphism. Show that f is an isomorphism if and only if $F(f)$ is an isomorphism.*

Solution. We know that if f is an isomorphism, then $F(f)$ is also an isomorphism without any special assumption concerning the functor F.

Suppose now that $F(f)\colon F(A) \to F(B)$ is an isomorphism. Then there exists a Δ-homomorphism $\gamma\colon F(B) \to F(A)$ such that

$$\gamma F(f) = i_{F(A)}$$

and $F(f)\gamma = i_{F(B)}$. Since F is fully faithful, there is a Λ-homomorphism $g\colon B \to A$ such that $F(g) = \gamma$. Hence $F(gf) = F(g)F(f) = \gamma F(f) = F(i_A)$ and $F(fg) = F(f)F(g) = F(f)\gamma = F(i_B)$. It follows, since F is fully faithful, that $gf = i_A$ and $fg = i_B$. Thus f is an isomorphism.

Exercise 5. *Let $F\colon \mathscr{C}_\Lambda \to \mathscr{C}_\Delta$ be an additive covariant functor and let $G\colon \mathscr{C}_\Delta \to \mathscr{C}_\Lambda$ be a functor in the reverse direction. Assume that GF resp. FG is naturally equivalent to the identity functor on $\mathscr{C}_\Lambda$ resp. $\mathscr{C}_\Delta$. Show that G is additive. (Thus F and G are inverse equivalences.)*

Solution. Let f,g belong to $\operatorname{Hom}_\Delta(A,B)$. Since the identity functor on $\mathscr{C}_\Delta$ is additive and FG is naturally equivalent to it, FG is also

130   POLYNOMIAL RINGS AND MATRIX RINGS

additive. Hence $FG(f+g) = FG(f) + FG(g) = F(G(f) + G(g))$ because F is additive as well. The discussion in the text shows that F is fully faithful. Consequently $G(f+g) = G(f) + G(g)$, that is, G is additive.

Exercise 6. *Let* $F: \mathscr{C}_\Lambda \to \mathscr{C}_\Delta$ *and* $G: \mathscr{C}_\Delta \to \mathscr{C}_\Lambda$ *be inverse equivalences. Show that if A is a Λ-module and K a Δ-module, then there exists an isomorphism*

$$\operatorname{Hom}_\Lambda(A, G(K)) \xrightarrow{\sim} \operatorname{Hom}_\Delta(F(A), K),$$

of abelian groups, which is natural for homomorphisms $A' \to A$ and $K \to K'$ in $\mathscr{C}_\Lambda$ and $\mathscr{C}_\Delta$ respectively.

Solution. We have a natural equivalence $\omega: FG \to I_\Delta$. If K is a Δ-module, then $\omega_K: FG(K) \xrightarrow{\sim} K$ induces an isomorphism

$$\operatorname{Hom}_\Delta(F(A), FG(K)) \xrightarrow{\sim} \operatorname{Hom}_\Delta(F(A), K).$$

Now, by Theorem 7, F is fully faithful and therefore it gives rise to an isomorphism $\operatorname{Hom}_\Lambda(A, G(K)) \xrightarrow{\sim} \operatorname{Hom}_\Delta(F(A), FG(K))$. By combining, we arrive at an isomorphism

$$\operatorname{Hom}_\Lambda(A, G(K)) \xrightarrow{\sim} \operatorname{Hom}_\Delta(F(A), K)$$

in which f in $\operatorname{Hom}_\Lambda(A, G(K))$ corresponds to $\omega_K F(f)$ in

$$\operatorname{Hom}_\Delta(F(A), K).$$

Suppose that $u: A' \to A$ in $\mathscr{C}_\Lambda$ and $v: K \to K'$ in $\mathscr{C}_\Delta$. Then, because $v\omega_K = \omega_{K'} FG(v)$, we have

$$\omega_{K'} F(G(v)fu) = \omega_{K'}(FG(v)) F(f) F(u) = v(\omega_K F(f)) F(u)$$

which shows that the diagram

$$
\begin{array}{ccc}
\operatorname{Hom}_\Lambda(A, G(K)) & \xrightarrow{\sim} & \operatorname{Hom}_\Delta(F(A), K) \\
\downarrow & & \downarrow \\
\operatorname{Hom}_\Lambda(A', G(K')) & \xrightarrow{\sim} & \operatorname{Hom}_\Delta(F(A'), K')
\end{array}
$$

is commutative. This completes the solution.

Exercise 7. *Let* $F: \mathscr{C}_\Lambda \to \mathscr{C}_\Delta$ *be an equivalence and let K be a Λ-module. Show that K is a generator of $\mathscr{C}_\Lambda$ if and only if $F(K)$ is a generator of $\mathscr{C}_\Delta$.*

Solution. Suppose first that $F(K)$ is a generator of $\mathscr{C}_\Delta$ and let A, B be Λ-modules. We wish to show that the natural Z-homomorphism

$\operatorname{Hom}_\Lambda(A, B) \to \operatorname{Hom}_Z(\operatorname{Hom}_\Lambda(K, A), \operatorname{Hom}_\Lambda(K, B))$ is a monomorphism. Assume that $f \in \operatorname{Hom}_\Lambda(A, B)$ and $\operatorname{Hom}_\Lambda(K, f) = 0$. Since the diagram

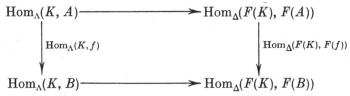

is commutative (the horizontal mappings are the Z-isomorphisms induced by F), it follows that $\operatorname{Hom}_\Lambda(F(K), F(f)) = 0$ whence $F(f) = 0$ because $F(K)$ is a generator. But F is fully faithful; consequently $f = 0$. This shows that the functor $\operatorname{Hom}_\Lambda(K, -)$ is faithful and hence that K is a generator.

Suppose now that K is a generator of $\mathscr{C}_\Lambda$ and let $G : \mathscr{C}_\Lambda \to \mathscr{C}_\Lambda$ be an equivalence inverse to F. Then $GF(K)$ is isomorphic to K and therefore $GF(K)$ is a generator of $\mathscr{C}_\Lambda$. By the first part, it follows that $F(K)$ is a generator of $\mathscr{C}_\Lambda$.

Exercise 8. *Let E be a non-zero injective Λ-module. Show that the following statements are equivalent:*

(a) *E has no non-trivial direct summands;*
(b) *E is the injective envelope of each of its non-zero submodules;*
(c) *every pair of non-zero submodules of E has a non-zero intersection;*
(d) *the non-units of $\operatorname{End}_\Lambda(E)$ form a two-sided ideal.*

Solution. *Assume* (a). Let A be a non-zero submodule of E. Then, by (Chapter 2, Theorem 17), E contains an injective envelope E' of A, and E', by (Chapter 2, Theorem 15), must be a direct summand of E. Consequently $E' = E$. Thus (a) implies (b).

Assume (b). Suppose that A and B are non-zero submodules of E. Then, by hypothesis, E is an injective envelope of B and therefore an essential extension of B. Consequently $A \cap B$ is non-zero and we have shown that (c) follows from (b).

Now assume (c). Let $h \in \operatorname{End}_\Lambda(E)$ and suppose that $\operatorname{Ker} h = 0$. Then $h(E) \subseteq E$ and, since $h(E)$ is isomorphic to E, it is injective. Accordingly, by (Chapter 2, Theorem 15), $E = h(E) \oplus B$ for a suitable submodule B of E. But $h(E) \neq 0$ and $h(E) \cap B = 0$. Consequently, by (c), $B = 0$ and therefore $h(E) = E$. Thus $h : E \to E$ is an automorphism, i.e. h is a unit of $\operatorname{End}_\Lambda(E)$. It follows that if $f \in \operatorname{End}_\Lambda(E)$, then f is a non-unit if and only if $\operatorname{Ker} f \neq 0$.

132   POLYNOMIAL RINGS AND MATRIX RINGS

Suppose that f_1 and f_2 are non-units of $\text{End}_\Lambda (E)$ and that ϕ is an arbitrary element of $\text{End}_\Lambda (E)$. Then $\text{Ker} f_1 \neq 0$, $\text{Ker} f_2 \neq 0$ hence, by (c), $\text{Ker} f_1 \cap \text{Ker} f_2 \neq 0$. It follows that $\text{Ker} (f_1 + f_2) \neq 0$ and therefore $f_1 + f_2$ is a non-unit. Again $\text{Ker} f_1 \subseteq \text{Ker} (\phi f_1)$ so we see that ϕf_1 is a non-unit as well. We must now show that $f_1 \phi$ is a non-unit. However if ϕ is a unit this is clear whereas if ϕ is a non-unit, then $\text{Ker} \phi \neq 0$ and $\text{Ker} \phi \subseteq \text{Ker} (f_1 \phi)$. Thus, in any event, $f_1 \phi$ is a non-unit. These remarks show that (d) is a consequence of (c).

Finally assume (d). Let $E = A_1 \oplus A_2$, where A_1, A_2 are submodules of E and let $\sigma_\nu : A_\nu \to E$ and $\pi_\nu : E \to A_\nu$ be the canonical homomorphisms. Then $\sigma_1 \pi_1 + \sigma_2 \pi_2 = i_E$ and therefore either $\sigma_1 \pi_1$ or $\sigma_2 \pi_2$ is a unit. For definiteness suppose that $\sigma_1 \pi_1$ is a unit. Then $E = \sigma_1 \pi_1(E) = A_1$ and therefore $A_2 = 0$. This completes the solution.

Exercise 9. *Show that if Λ and Δ are anti-isomorphic, then $\mathscr{C}_\Lambda^L$ and $\mathscr{C}_\Delta^R$ are equivalent. Deduce that $\mathscr{C}_\Lambda^L$ and $\mathscr{C}_{M_n(\Lambda)}^L$ are equivalent, where n is an arbitrary positive integer.*

Solution. Let $\phi : \Lambda \to \Delta$ be an anti-isomorphism and let A be a left Λ-module. For $a \in A$ and $\delta \in \Delta$ define $a\delta$ by $a\delta = \lambda a$, where λ denotes the element of Λ determined by $\phi(\lambda) = \delta$. This turns A into a right Δ-module. We use $F(A)$ to designate A considered as a right Δ-module.

Let A, A' be left Λ-modules. A mapping $f : A \to A'$ is a Λ-homomorphism if and only if, considered as a mapping of $F(A)$ into $F(A')$, it is a Δ-homomorphism. When $f : A \to A'$ is a Λ-homomorphism, the corresponding Δ-homomorphism $F(A) \to F(A')$ will be denoted by $F(f)$. This provides an additive covariant functor $F : \mathscr{C}_\Lambda^L \to \mathscr{C}_\Delta^R$ which is clearly an equivalence.

Let Λ^* be the ring opposite to Λ. Then we have a mapping

$$\psi : M_n(\Lambda) \to M_n(\Lambda^*)$$

in which each matrix in $M_n(\Lambda)$ is converted into its transpose. It is easily verified that this is an anti-isomorphism.

By the first part $\mathscr{C}_\Lambda^L$ is equivalent to $\mathscr{C}_{\Lambda^*}^R$ and $\mathscr{C}_{M_n(\Lambda)}^L$ to $\mathscr{C}_{M_n(\Lambda^*)}^R$. Since, by Theorem 16, $\mathscr{C}_{\Lambda^*}^R$ and $\mathscr{C}_{M_n(\Lambda^*)}^R$ are equivalent, the desired result follows.

Exercise 10. *The ring Λ is the direct sum of the rings $\Lambda_1, \Lambda_2, ..., \Lambda_s$. Show that $r.\text{GD}(\Lambda) = \max_{1 \leqslant \mu \leqslant s} r.\text{GD}(\Lambda_\mu)$.*

Solution. Let $\pi_\mu : \Lambda \to \Lambda_\mu$ be the canonical projection. This is a surjective ring-homomorphism and if we consider Λ_μ as a right Λ-module, then it is a direct summand of Λ and hence Λ-projective. Thus $r.\mathrm{Pd}_\Lambda(\Lambda_\mu) \leqslant 0$. It follows, from (Chapter 3, Exercise 12), that if U is a right Λ_μ-module, then $r.\mathrm{Pd}_\Lambda(U) \leqslant r.\mathrm{Pd}_{\Lambda_\mu}(U)$. In particular if U is Λ_μ-projective, then it is Λ-projective.

Put $e_\mu = \pi_\mu(1)$. Then e_μ belongs to the centre of Λ,

$$1 = e_1 + e_2 + \ldots + e_s, \quad e_\mu^2 = e_\mu, \quad e_\mu e_\nu = 0$$

when $\mu \neq \nu$, and $\Lambda_\mu = \Lambda e_\mu$. Let M be a Λ-module. Then Me_μ is both a Λ-module and a module over the ring $\Lambda_\mu = \Lambda e_\mu$ the two structures being connected through the homomorphism π_μ in the usual way. Now if $M \approx \underset{i \in I}{\oplus} M_i$ in the category $\mathscr{C}_\Lambda^R$, then $Me_\mu \approx \underset{i \in I}{\oplus} M_i e_\mu$ in $\mathscr{C}_{\Lambda_\mu}^R$. It follows that if F is a free right Λ-module, then Fe_μ is a free right Λ_μ-module.

Let U be a right Λ_μ-module. *We claim that U is Λ_μ-projective if and only if it is Λ-projective.* For suppose that U is Λ-projective, then in $\mathscr{C}_\Lambda^R$ we have a relation $U \oplus P = F$, where P is Λ-projective and F is Λ-free. Hence $Ue_\mu \oplus Pe_\mu = Fe_\mu$ whence $Ue_\mu = U$ is Λ_μ-projective. This establishes our claim because the reverse implication has already been established.

Once again let U be a right Λ_μ-module. *This time we claim that*

$$r.\mathrm{Pd}_\Lambda(U) = r.\mathrm{Pd}_{\Lambda_\mu}(U).$$

Indeed it is sufficient to show that

$$r.\mathrm{Pd}_{\Lambda_\mu}(U) \leqslant r.\mathrm{Pd}_\Lambda(U)$$

and for this we may suppose that $r.\mathrm{Pd}_\Lambda(U) = n$, where $0 \leqslant n < \infty$. To this end construct, in $\mathscr{C}_{\Lambda_\mu}^R$, an exact sequence

$$0 \to A_n \to P_{n-1} \to \ldots \to P_1 \to P_0 \to U \to 0,$$

where P_i is Λ_μ-projective and hence also Λ-projective. Then, because $r.\mathrm{Pd}_\Lambda(U) = n$, the Λ_μ-module A_n is Λ-projective and therefore also Λ_μ-projective. Hence $r.\mathrm{Pd}_{\Lambda_\mu}(U) \leqslant n$ and our claim follows.

We now see that $r.\mathrm{GD}(\Lambda_\mu) \leqslant r.\mathrm{GD}(\Lambda)$ and therefore

$$\max_{1 \leqslant \mu \leqslant s} r.\mathrm{GD}(\Lambda_\mu) \leqslant r.\mathrm{GD}(\Lambda).$$

Finally let M be a right Λ-module. Then

$$M = Me_1 \oplus Me_2 \oplus \ldots \oplus Me_s.$$

Accordingly, by (Chapter 3, Exercise 10),

$$r.\mathrm{Pd}_\Lambda\,(M) = \max_{1 \leqslant \mu \leqslant s} r.\mathrm{Pd}_\Lambda\,(Me_\mu)$$

$$= \max_{1 \leqslant \mu \leqslant s} r.\mathrm{Pd}_{\Lambda_\mu}\,(Me_\mu)$$

$$\leqslant \max_{1 \leqslant \mu \leqslant s} r.\mathrm{GD}\,(\Lambda_\mu).$$

Thus $\qquad\qquad r.\mathrm{GD}\,(\Lambda) \leqslant \max_{1 \leqslant \mu \leqslant s} r.\mathrm{GD}\,(\Lambda_\mu)$

and now the solution is complete.

5

DUALITY

5.1 General remarks

As usual, Λ denotes a ring with identity element and $\mathscr{C}_\Lambda^L$ resp. $\mathscr{C}_\Lambda^R$ denotes the category of left resp. right Λ-modules. When we do not need to specify in which of these two categories we are working, then we shall use $\mathscr{C}_\Lambda$. Now Λ itself can be regarded as either a left Λ-module or as a right Λ-module. When considered as a left Λ-module we shall use the symbol Λ_l and when we wish to regard it as a right Λ-module it will be designated by Λ_r.

Our main purpose will be to study $\mathrm{Hom}_\Lambda (-, \Lambda)$. This, as we have noted earlier,† may be regarded as a left exact contravariant functor taking left Λ-modules into right Λ-modules or, equally well, as a left exact contravariant functor from right Λ-modules to left Λ-modules. Thus if A is a Λ-module, then $A^* = \mathrm{Hom}_\Lambda (A, \Lambda)$ is also a Λ-module. However A^* is of the opposite type to A, i.e. it has the elements of Λ operating on the other side. We shall refer to A^* as the *dual* of A.

The prototype for our duality theory is the theory of finite-dimensional vector spaces. This theory does not extend to spaces where the dimension is arbitrary. Hence in our generalized theory we must expect to have to limit ourselves to the consideration of finitely generated modules. It is therefore highly advantageous to be in a situation where a submodule of a finitely generated module is always finitely generated, and this is tantamount to requiring that the ground-ring Λ be Noetherian.‡ We shall therefore devote an initial section to a survey of the basic facts concerning Noetherian rings and modules and it will be covenient, at the same time, to draw attention to the almost parallel Artinian theory because this will be needed later.

† See Chapter 2, Example 2.
‡ See the next section for the definition.

5.2 Noetherian and Artinian conditions

Theorem 1. *Let M be a Λ-module. Then the following two statements are equivalent:*

(a) *every non-empty set of submodules of M contains at least one maximal member;*

(b) *given an infinite increasing sequence $K_1 \subseteq K_2 \subseteq K_3 \subseteq \ldots$ of submodules of M, there exists a positive integer p such that $K_s = K_p$ for all $s \geqslant p$.*

Remark. When (a) holds we say that M satisfies the *maximal condition* for submodules, and when (b) holds that M satisfies the *ascending chain condition* for submodules. Thus the theorem asserts that the two conditions are equivalent. Another way of describing the ascending chain condition is to say that every infinite increasing sequence of submodules *terminates*.

Proof. Assume (a) and let $K_1 \subseteq K_2 \subseteq K_3 \subseteq \ldots$ be an increasing sequence of submodules of M. Then the family $\{K_\mu\}_{\mu \geqslant 1}$ contains a maximal member K_p say. Evidently $K_s = K_p$ for all $s \geqslant p$.

Now assume (b) and suppose that Ω is a non-empty set of submodules of M. We shall suppose that Ω has no maximal member and derive a contradiction. With this hypothesis, if $K \in \Omega$, then there exists $K' \in \Omega$ such that $K \subset K'$. (Remember $\subset$ means strict inclusion.) It follows that, because Ω is non-empty, we can generate an infinite strictly increasing sequence

$$K_1 \subset K_2 \subset K_3 \subset \ldots$$

of submodules of Ω. However this violates (b).

Definition. *Let M be a Λ-module. If M satisfies the two equivalent conditions of Theorem 1, then M is said to be a 'Noetherian' Λ-module.*

Definition. *We say that the ring Λ is 'left resp. right Noetherian' if Λ_l resp. Λ_r is a Noetherian Λ-module.*

Theorem 2. *Let M be a Λ-module. Then the following two statements are equivalent:*

(a) *every non-empty set of submodules of M contains at least one minimal member;*

(b) *every infinite decreasing sequence $K_1 \supseteq K_2 \supseteq K_3 \supseteq \ldots$ of submodules of M terminates, that is becomes constant from some point onwards.*

The proof of this result is so similar to that of Theorem 1 that we shall omit it. When (a) holds M is said to satisfy the *minimal condition* for submodules. The condition described in (b) is called the *descending chain condition*.

Definition. *A Λ-module M which satisfies the two equivalent conditions of Theorem 2 is called an 'Artinian' Λ-module.*

Definition. *The ring Λ is said to be 'left resp. right Artinian' if Λ_l resp. Λ_r is an Artinian Λ-module.*

Theorem 3.† *A Λ-module M is Noetherian if and only if each of its submodules is finitely generated.*

Proof. First suppose that M is Noetherian and let K be a submodule of M. The set of finitely generated submodules of K is not empty and therefore it possesses a maximal member K' say. To show that K is finitely generated it is enough to prove that $K = K'$. Evidently $K' \subseteq K$. Let $y \in K$. Then there is a finitely generated submodule K'', of K, containing both K' and y. By the choice of K', we must have $K' = K''$ and therefore $y \in K'$. This shows that $K = K'$.

Now suppose that all the submodules of M are finitely generated. Let $N_1 \subseteq N_2 \subseteq N_3 \subseteq \ldots$ be an infinite increasing sequence of submodules of M. To complete the proof it is enough to show that the sequence terminates. Put $N = \bigcup_{\nu=1}^{\infty} N_\nu$. An easy verification shows that N is a submodule of M. It is therefore finitely generated say by $x_1, x_2, \ldots, x_q$. Choose p so that N_p contains all the x_i. Then $N \subseteq N_p \subseteq N$ and therefore $N_s = N_p$ for $s \geqslant p$.

Corollary. *The ring Λ is left resp. right Noetherian if and only if each left resp. right ideal is finitely generated.*

The Noetherian condition has interesting connections with the theory of injectives. The next two exercises give examples of this.

Exercise 1. *Let Λ be left Noetherian. Show that every direct sum of injective left Λ-modules is injective.*

Exercise 2.* *Given that every countable direct sum of injective left Λ-modules is again injective, deduce that Λ is left Noetherian.*

Theorem 4. *Let $0 \to A \to B \to C \to 0$ be an exact sequence of Λ-modules. Then B is Noetherian if and only if A and C are both Noetherian. In*

---

† There is a counterpart to this result in the theory of Artinian modules, but we shall not say anything about it here. See, however, (25) Theorem 3.21.

particular, all submodules and factor modules of a Noetherian module are themselves Noetherian.

Proof. We shall assume that A and C are Noetherian and deduce that B is Noetherian. (The converse is trivial.) For this we may assume that A is a submodule of B.

Let $K_1 \subseteq K_2 \subseteq K_3 \subseteq \ldots$ be an increasing sequence of submodules of B. Since A is Noetherian, the sequence

$$(K_1 \cap A) \subseteq (K_2 \cap A) \subseteq (K_3 \cap A) \subseteq \ldots$$

terminates and, since C is Noetherian, the same is true of

$$(K_1 + A) \subseteq (K_2 + A) \subseteq (K_3 + A) \subseteq \ldots.$$

Hence there exists an integer p such that $K_p \cap A = K_s \cap A$ and $K_p + A = K_s + A$ for all $s \geqslant p$.

Suppose that $s \geqslant p$. Then

$$K_s = K_s \cap (K_s + A) = K_s \cap (K_p + A) = K_p + (K_s \cap A)$$

because $K_p \subseteq K_s$. Accordingly $K_s = K_p + (K_p \cap A) = K_p$ and we have shown that the sequence $K_1 \subseteq K_2 \subseteq K_3 \subseteq \ldots$ terminates. This proves the theorem.

Corollary. *Let $M_1, M_2, \ldots, M_p$ be Noetherian Λ-modules. Then*

$$M_1 \oplus M_2 \oplus \ldots \oplus M_p$$

is also Noetherian.

Proof. We may suppose that $p = 2$. In this case the assertion follows at once from the canonical exact sequence

$$0 \to M_1 \to M_1 \oplus M_2 \to M_2 \to 0$$

and Theorem 4.

Exercise 3.* *Suppose that A is a Noetherian left Λ-module and that x is an indeterminate. Show that the polynomial module† $A[x]$ is Noetherian when considered as a left module with respect to the polynomial ring $\Lambda[x]$.*

This incorporates the famous *Basis Theorem* of Hilbert. The result is very useful for constructing examples of Noetherian rings. It occurs here as an exercise rather than as a theorem because it will not play a significant role in the Duality Theory to which this chapter is devoted.

† See section (4.2).

Theorem 5. *Suppose that Λ is left Noetherian and let M be a left Λ-module. Then M is Noetherian if and only if it is finitely generated.*

Proof. If M is Noetherian, then it is finitely generated by Theorem 3. Now assume that M is finitely generated. We can construct an exact sequence $0 \to K \to F \to M \to 0$, where F has the form

$$\Lambda_l \oplus \Lambda_l \oplus \dots \oplus \Lambda_l.$$

That F is Noetherian follows from Theorem 4 Cor. and now an application of Theorem 4 shows that M is Noetherian as well.

Corollary. *Suppose that Λ is left Noetherian and M is a finitely generated left Λ-module. Then every submodule of M is finitely generated.*

This follows by combining Theorem 5 with Theorem 3.

Exercise 4.* *Give an example of a ring which is left Noetherian but not right Noetherian.*

Theorem 6. *Let $0 \to A \to B \to C \to 0$ be an exact sequence of Λ-modules. Then B is Artinian if and only if A and C are both Artinian. In particular all submodules and factor modules of an Artinian module are themselves Artinian.*

A proof can be obtained by making minor modifications to the arguments used to establish Theorem 4. The proof of the corollary to Theorem 4 can also be adapted to prove the following

Corollary. *Let $M_1, M_2, \dots, M_p$ be Artinian Λ-modules. Then*

$$M_1 \oplus M_2 \oplus \dots \oplus M_p$$

is also Artinian.

Exercise 5.* *Give an example of a ring which is left Artinian but not right Artinian.*

Theorem 7. *Let Λ be left Artinian and M a finitely generated left Λ-module. Then M is Artinian.*

Proof. M is a homomorphic image of a direct sum $\Lambda_l \oplus \Lambda_l \oplus \dots \oplus \Lambda_l$. The assertion now follows from Theorem 6 and its corollary.

5.3 Preliminaries concerning duality

For the present we shall make no special assumptions concerning the ring Λ. Let A be a Λ-module and put

$$A^* = \operatorname{Hom}_\Lambda (A, \Lambda). \tag{5.3.1}$$

Then A^* is also a Λ-module but of the opposite type to A. It is called the *dual* of A and $\mathrm{Hom}_{\Lambda}(-, \Lambda)$ is called the *duality functor*. (Strictly speaking there are two duality functors, one from $\mathscr{C}_{\Lambda}^L$ to $\mathscr{C}_{\Lambda}^R$ and the other from $\mathscr{C}_{\Lambda}^R$ to $\mathscr{C}_{\Lambda}^L$. However either the context will make clear which it is that we have in mind or our remarks will apply in both cases.) Observe that if $f \in A^*$, $\lambda \in \Lambda$ and A is a *left* Λ-module, then $(f\lambda)(a) = f(a)\lambda$ for a in A, whereas if A is a *right* Λ-module, then $(\lambda f)(a) = \lambda f(a)$. Of course, isomorphic modules have isomorphic duals.

Consider $\Lambda_l^* = \mathrm{Hom}_{\Lambda}(\Lambda_l, \Lambda)$. This is a right Λ-module. We have in fact a Λ-isomorphism

$$\Lambda_l^* \approx \Lambda_r \qquad (5.3.2)$$

in which f in Λ_l^* corresponds to $f(1)$ in Λ_r. This isomorphism will sometimes be used to identify Λ_l^* with Λ_r. There is also a similar isomorphism

$$\Lambda_r^* \approx \Lambda_l \qquad (5.3.3)$$

in which g in Λ_r^* corresponds to $g(1)$ in Λ_l. This allows us to identify Λ_r^* with Λ_l.

Let F be a free Λ-module with a finite base $e_1, e_2, \ldots, e_n$. For each i $(1 \leqslant i \leqslant n)$ define a Λ-homomorphism $\phi_i : F \to \Lambda$ by $\phi_i(e_j) = \delta_{ij}$, where δ_{ij} is the Kronecker symbol. Then $\phi_i \in F^*$. *We claim that F^* is a free Λ-module having $\phi_1, \phi_2, \ldots, \phi_n$ as a base.* In establishing this we shall assume for definiteness that F is a left Λ-module and therefore F^* is a right Λ-module. Let $\lambda_1, \lambda_2, \ldots, \lambda_n$ belong to Λ. Then

$$(\phi_1 \lambda_1 + \phi_2 \lambda_2 + \ldots + \phi_n \lambda_n)(e_j) = \sum_{i=1}^{n} (\phi_i \lambda_i)(e_i)$$

$$= \sum_{i=1}^{n} \phi_i(e_j) \lambda_i$$

and therefore $(\phi_1 \lambda_1 + \ldots + \phi_n \lambda_n)(e_j) = \lambda_j$. Our claim is now obvious. It is usual to refer to $\phi_1, \phi_2, \ldots, \phi_n$ as the *base of F^* dual to the base* $e_1, e_2, \ldots, e_n$ *of F*.

Exercise 6. *Show that if P is a projective Λ-module which can be generated by n elements, where $0 \leqslant n < \infty$, then the dual module P^* is also Λ-projective and it too can be generated by n elements.*

Exercise 7. *Let Λ be a right Noetherian ring. Show that the dual of any finitely generated left Λ-module is also finitely generated.*

Let A be a Λ-module with dual A^*. Then A^* itself has a dual A^{**}.

This is called the *double dual* or *bidual* of A. Evidently A^{**} is an additive *covariant* functor of A. Suppose that $a \in A$ and

$$f \in A^* = \mathrm{Hom}_\Lambda (A, \Lambda).$$

Define a mapping $\phi_a : A^* \to \Lambda$ by $\phi_a(f) = f(a)$. Obviously

$$\phi_a(f_1 + f_2) = \phi_a(f_1) + \phi_a(f_2).$$

Let us assume that A is a left Λ-module and that $\lambda \in \Lambda$. Then

$$\phi_a(f\lambda) = f(a)\lambda = \phi_a(f)\lambda.$$

Thus ϕ_a is a Λ-homomorphism, i.e. $\phi_a \in A^{**}$ and we arrive at the same conclusion if A belongs to $\mathscr{C}_\Lambda^R$.

Consider the mapping
$$\delta_A : A \to A^{**} \tag{5.3.4}$$

defined by $\delta_A(a) = \phi_a$ so that, if $f \in A^*$, then

$$(\delta_A(a))(f) = f(a). \tag{5.3.5}$$

A straightforward verification shows that δ_A is a Λ-homomorphism. We shall refer to δ_A as the *canonical homomorphism* of A into its bidual. Let $u : A \to B$ be a Λ-homomorphism. This will induce a Λ-homomorphism $u^* : B^* \to A^*$ and this in turn will give rise to a homomorphism $u^{**} : A^{**} \to B^{**}$. It is easy to check that the diagram

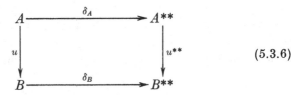

$$\tag{5.3.6}$$

is commutative. Accordingly the homomorphisms δ_A constitute a natural transformation of the identity functor on $\mathscr{C}_\Lambda$ into the bidual functor.

Definition. *The Λ-module A is called 'reflexive' if δ_A is an isomorphism; it is called 'semi-reflexive' if δ_A is a monomorphism.*

Semi-reflexive modules are also known as *torsionless* modules. The reason for this name is explained by the next theorem. Observe that if a Λ-module A is reflexive resp. semi-reflexive, then any module isomorphic to A is also reflexive resp. semi-reflexive. The lemma which follows is just a useful restatement of the definition of a semi-reflexive module.

Lemma 1. *Let A be a Λ-module. Then A is semi-reflexive if and only if for each $a \in A, a \neq 0$, there exists $f \in \mathrm{Hom}_\Lambda(A, \Lambda)$ such that $f(a) \neq 0$.*

We shall now illustrate some of our concepts by means of an excursion into the theory of integral domains. Let R be an integral domain and M an R-module. An element $x \in M$ is called a *torsion element* if $cx = 0$ for some $c \in R, c \neq 0$. The torsion elements form a submodule N which is called the *torsion submodule* of M. If $N = 0$, then M is said to be *torsion free*. If, however, $N = M$ then M is said to be a *torsion module*. For example N itself is a torsion module and M/N is torsion free.

Theorem 8. *Let R be an integral domain and A a finitely generated R-module. Then A is semi-reflexive if and only if it is torsion free.*

Proof. Assume first that A is a semi-reflexive. Let $a \in A$, $r \in R$ with $a \neq 0, r \neq 0$. By Lemma 1, there exists $f \in \mathrm{Hom}_R(A, R)$ such that $f(a) \neq 0$. Then $rf(a) \neq 0$, because R is an integral domain, and therefore $f(ra) \neq 0$. It follows that $ra \neq 0$ and this proves that A is torsion free.

Now assume that A is torsion free. We wish to show that A is semi-reflexive so we may suppose that $A \neq 0$. Let $a_1, a_2, ..., a_n$ generate A and from among these select a maximal subset that is linearly independent with respect to R. We may take it that $a_1, a_2, ..., a_s$ are the members of the selected subset. It is now possible to find $c \in R$, $c \neq 0$ so that $cA \subseteq Ra_1 + Ra_2 + ... + Ra_s$. Note that $Ra_1 + Ra_2 + ... + Ra_s$ is a free R-module with $a_1, a_2, ..., a_s$ as base. Denote by ϕ_ν the R-homomorphism of $Ra_1 + Ra_2 + ... + Ra_s$ into R which maps

$$r_1 a_1 + r_2 a_2 + ... + r_s a_s$$

into r_ν.

To complete the proof suppose that $a \in A, a \neq 0$. Then

$$ca = r_1 a_1 + r_2 a_2 + ... + r_s a_s,$$

where the r_i are in R and we can select one of them, say r_ν, so that $r_\nu \neq 0$. (This is because A is torsion free.) Define an R-homomorphism $f: A \to R$ by $f(\alpha) = \phi_\nu(c\alpha)$. Then $f(a) = r_\nu$ and therefore $f(a) \neq 0$. That A is semi-reflexive now follows from Lemma 1.

Let R be an integral domain and F its quotient field. If I is an R-submodule of F, then we put

$$I^{-1} = \{\alpha | \alpha \in F \text{ and } \alpha I \subseteq R\}.$$

It is easily verified that I^{-1} is also an R-submodule of F.

Theorem 9. *Let R be an integral domain and $I \neq (0)$ an ideal of R. Then $I^* = \text{Hom}_R(I, R)$ and I^{-1} are isomorphic R-modules.*

Proof. Suppose that $a, b \in I$ and $a \neq 0, b \neq 0$. Then if $f \in I^*$ we have $f(a)/a = f(b)/b$. Thus the element $q = f(a)/a$, of F, is independent of the choice of the non-zero element a of I. Further if $c \in I$, then $qc = f(c) \in R$ which shows that $q \in I^{-1}$. Accordingly there exists an R-homomorphism $I^* \to I^{-1}$ in which f is mapped into q. Obviously this is a monomorphism.

Assume that $q' \in I^{-1}$. We can define an R-homomorphism $\phi : I \to R$ by $\phi(c) = cq'$. Then $\phi \in I^*$ and its image in I^{-1} is q'. These remarks prove that the mapping $I^* \to I^{-1}$ is an isomorphism of R-modules.

Exercise 8. *Let R be an integral domain with quotient field F and $I \neq (0)$ an ideal of R. Show that I is reflexive if and only if $I = (I^{-1})^{-1}$.*

We shall now leave the discussion of integral domains and return to the general theory.

Exercise 9. *Show that a submodule of a semi-reflexive module is semi-reflexive and that a direct summand of a reflexive module is reflexive.*

Theorem 10. *A finitely generated projective module is reflexive. In particular, a free module with a finite base is reflexive.*

Proof. Let F be a free module with a base $e_1, e_2, \ldots, e_n$ and let

$$\phi_1, \phi_2, \ldots, \phi_n$$

be the base of F^* dual to $e_1, e_2, \ldots, e_n$. In F^{**} let $\psi_1, \psi_2, \ldots, \psi_n$ be the base dual to $\phi_1, \phi_2, \ldots, \phi_n$. Then $\delta_F(e_i)$ and ψ_i both belong to

$$\text{Hom}_\Lambda(F^*, \Lambda)$$

and $\quad (\delta_F(e_i))(\phi_j) = \phi_j(e_i) = \delta_{ji} = \psi_i(\phi_j).$

It follows that $\delta_F(e_i) = \psi_i$ and hence that $\delta_F : F \to F^{**}$ is an isomorphism taking the base $e_1, e_2, \ldots, e_n$ of F into the base $\psi_1, \psi_2, \ldots, \psi_n$ of F^{**}. In particular F is reflexive.

Now let P be a projective module which can be generated by n elements. Then, by (Chapter 2, Theorem 7 Cor.), P is a direct summand of a free module with a base of n elements. That P is reflexive now follows from Exercise 9.

Let A be a Λ-module. If we apply the duality functor to the Λ-homomorphism $\delta_A : A \to A^{**}$, then we obtain a Λ-homomorphism

$$\delta_A^* : A^{***} \to A^*. \tag{5.3.7}$$

In addition we have the canonical homomorphism

$$\delta_{A^*}: A^* \to A^{***}. \tag{5.3.8}$$

We claim that

$$\delta_A^* \delta_{A^*} = i_{A^*}. \tag{5.3.9}$$

Exercise 10. *Establish the relation* $\delta_A^* \delta_{A^*} = i_{A^*}$.

As a simple application of (5.3.9) we prove

Theorem 11. *The dual of an arbitrary module is semi-reflexive. The dual of a reflexive module is reflexive.*

Proof. Let A be a Λ-module. By (5.3.9), δ_{A^*} is a monomorphism and therefore A^* is semi-reflexive. Now assume that A is reflexive. Then δ_A is an isomorphism and therefore so too is δ_A^*. By (5.3.9),

$$\delta_{A^*} = (\delta_A^*)^{-1}.$$

Accordingly δ_{A^*} is an isomorphism and the proof is complete.

5.4 Annihilators

Let A be a Λ-module and A^* its dual. Suppose that K is a submodule of A and denote by K^0 the set consisting of all f in A^* such that $f(x) = 0$ for every x in K. Then K^0 is a submodule of A^*. We shall call it the *annihilator of K in A^**. Now assume that M is a submodule of A^* and let M^0 consist of all $x \in A$ such that $f(x) = 0$ for every $f \in M$. This is a submodule of A. It will be called the *annihilator of M in A*. (Note that M also has an annihilator in A^{**}, but for the present this will not concern us.) The next lemma summarizes some of the elementary properties of annihilators. In it K resp. M denotes a submodule of A resp. A^* and K^0 resp. M^0 its annihilator in A^* resp. A.

Lemma 2. *The annihilator operators have the following properties*:
 (a) *if* $K_1 \subseteq K_2$ *then* $K_2^0 \subseteq K_1^0$, *and if* $M_1 \subseteq M_2$ *then* $M_2^0 \subseteq M_1^0$;
 (b) $K \subseteq K^{00}$ *and* $M \subseteq M^{00}$;
 (c) $K^0 = K^{000}$ *and* $M^0 = M^{000}$;
 (d) $(K_1 + K_2)^0 = K_1^0 \cap K_2^0$ *and* $(M_1 + M_2)^0 = M_1^0 \cap M_2^0$.

Proof. The assertions (a), (b) and (d) are trivial. From (a) and (b) we deduce that $K^{000} \subseteq K^0$. However if in $M \subseteq M^{00}$ we replace M by K^0, then we obtain the opposite inequality. Thus $K^0 = K^{000}$ and $M^0 = M^{000}$ similarly.

We saw in (5.3.2) that we have a Λ-isomorphism $\Lambda_l^* \approx \Lambda_r$ in which f in Λ_l^* corresponds to $f(1)$ in Λ_r. Suppose that $\lambda \in \Lambda_l$ and $f \in \Lambda_l^*$.

Then $f(\lambda) = \lambda f(1)$. Assume that I is a left ideal of Λ, that is a sub-module of Λ_l, and let I^0 be its annihilator in Λ_l^*. Then $f \in I^0$ if and only if $f(\lambda) = 0$ for all $\lambda \in I$, i.e. if and only if $If(1) = 0$. Thus *if we identify Λ_l^* with Λ_r, then the annihilator I^0 (of I in Λ_l^*) is simply the right annihilator of the left ideal I in the obvious sense.*

Next consider a right ideal H of Λ, that is a submodule of Λ_r. This will have an annihilator H^0 in Λ_r^*. However, as we saw in (5.3.3), we have a Λ-isomorphism $\Lambda_r^* \approx \Lambda_l$. *If this is used to identify Λ_r^* with Λ_l, then H^0 becomes the left annihilator of the right ideal H in the obvious sense.*

The observations of the last two paragraphs may be combined. Let I be a left ideal of Λ and consider it as a submodule of Λ_l. *Then I^{00} is the left annihilator of the right annihilator of I.* On the other hand, if H is a right ideal and we regard it as a submodule of Λ_r, *then H^{00} is the right annihilator of the left annihilator of H.*

Exercise 11. *Let $K_1, K_2, ..., K_n$ be submodules of a Λ-module A and suppose that A is their direct sum. Put*

$$U_i = K_1 + ... + K_{i-1} + K_{i+1} + ... + K_n$$

so that U_i is a submodule of A. Show that $A^* = U_1^0 \oplus U_2^0 \oplus ... \oplus U_n^0$ and that U_i^0 and K_i^* are isomorphic Λ-modules.

Definition. *A submodule K of the Λ-module A is said to be 'closed in A' if the annihilator, in A, of the annihilator of K in A^* is K itself.*

Thus K is closed in A if and only if $K^{00} = K$ in which case we shall also say that K is a *closed submodule* of A. The slightly cumbersome wording is used to prevent a possible source of confusion. Suppose that M is a submodule of A^*. To determine whether M is closed in A^* we have to consider the annihilator, in A^*, of the annihilator of M in A^{**} (not in A). However, as we shall now show this difficulty disappears when A is reflexive.

Lemma 3. *Let A be a reflexive Λ-module, M a submodule of its dual A^*, and M^0 the annihilator of M in A. Then $\delta_A(M^0)$ is the annihilator $\overline{M^0}$ (say) of M in A^{**}. Further M^0 and $\overline{M^0}$ have the same annihilator in A^* and therefore M is closed in A^* if and only if $M = M^{00}$.*

Proof. For $a \in A$ put $\phi_a = \delta_A(a)$. Then $\phi_a \in \mathrm{Hom}_\Lambda(A^*, \Lambda)$ and $\phi_a(f) = f(a)$ for all $f \in A^*$. Thus $a \in M^0$ if and only if $\phi_a(f) = 0$ for all $f \in M$ that is if and only if $\phi_a \in \overline{M^0}$. Since $\delta_A : A \to A^{**}$ is an isomorphism this proves the first assertion.

Let $g \in A^*$. Then $g(a) = 0$ for all $a \in M^0$ if and only if $\phi_a(g) = 0$ for all $a \in M^0$. Accordingly $g(a) = 0$ for all $a \in M^0$ if and only if $\theta(g) = 0$ for all $\theta \in \overline{M^0}$. This proves the second assertion and establishes the lemma.

We recall that the duality functor is left exact.

Theorem 12. *Let* $0 \to K \xrightarrow{j} A \xrightarrow{\phi} B \to 0$ *be an exact sequence of Λ-modules (where j is an inclusion mapping) and let* $0 \to B^* \to A^* \to K^*$ *be the exact sequence obtained by applying the duality functor. Then B is semi-reflexive if and only if K is closed in A. In any event B^*, considered as a submodule of A^*, coincides with the annihilator, K^0, of K in A^*.*

Proof. Let f belong to $B^* = \mathrm{Hom}_\Lambda (B, \Lambda)$. Then the homomorphism $B^* \to A^*$ takes f into $f\phi$. But the members of $A^* = \mathrm{Hom}_\Lambda (A, \Lambda)$ that have the form $f\phi$ are just those that vanish on K, i.e. they are precisely the members of K^0. This proves the final assertion.

Suppose that $K = K^{00}$. Let $b \in B, b \neq 0$ and choose $a \in A$ so that $\phi(a) = b$. Then $a \notin K$ and therefore $a \notin K^{00}$. Accordingly there exists $g \in K^0$ such that $g(a) \neq 0$. But, as is shown by the remarks of the last paragraph, $g = f\phi$ for some $f \in \mathrm{Hom}_\Lambda (B, \Lambda)$. It follows that $f(b) \neq 0$. We may now conclude that B is semi-reflexive by appealing to Lemma 1.

Finally suppose that B is semi-reflexive and let $a \in K^{00}$. Since $K \subseteq K^{00}$, the proof will be complete if we show that a belongs to K. Suppose that $a \notin K$. Then $\phi(a) \neq 0$ and therefore, by Lemma 1, there exists $f \in \mathrm{Hom}_\Lambda (B, \Lambda)$ such that $f\phi(a) \neq 0$. But we know that $f\phi \in K^0$ and by hypothesis $a \in K^{00}$. Consequently $f\phi(a) = 0$ and with this contradiction the proof is complete.

Theorem 13. *Let K be a left Λ-module and $n \geqslant 0$ an integer. Then the following statements are equivalent:*

(1) *K is a closed submodule of a free module with a base of n elements;*

(2) *K is isomorphic to the dual of a right Λ-module which can be generated by n elements.*

Proof. *Assume* (1). Then K is a closed submodule of a free module F with a base of n elements. Let K^0 be the annihilator of K in F^*. Then we have an exact sequence $0 \to K^0 \to F^* \to B \to 0$. Now F^* is also free and it too has a base of n elements. Consequently B is a right Λ-module which can be generated by n elements. By applying the duality functor, we arrive at an exact sequence

$$0 \to B^* \to F^{**} \to (K^0)^*$$

and, by Theorem 12, B^* is isomorphic to the annihilator of K^0 in F^{**}. But F is reflexive (Theorem 10). Consequently, by Lemma 3, B^* is isomorphic to K^{00} and $K^{00} = K$ because K is closed in F.

Assume (2). We may then suppose that $K = B^*$, where B is a right Λ-module generated by n elements. Construct an exact sequence $0 \to Q \to G \to B \to 0$, where G is a free module with a base of n elements and Q is a submodule of G. On applying the duality functor we obtain an exact sequence $0 \to B^* \to G^* \to Q^*$, where G^* is a free module with a base of n elements. It is now enough to show that B^*, considered as a submodule of G^*, is closed in G^*. This, however, follows from the next lemma.

Lemma 4. *Let* $A \overset{\phi}{\to} B \to 0$ *be an exact sequence of* Λ-*modules. Then the dual sequence* $0 \to B^* \overset{\phi^*}{\to} A^*$ *is also exact and* B^*, *considered as a submodule of* A^*, *is closed in* A^*.

Proof. Put $K = \operatorname{Ker} \phi$. Then $0 \to K \overset{j}{\to} A \overset{\phi}{\to} B \to 0$ is an exact sequence, where j is an inclusion mapping, and therefore $0 \to B^* \overset{\phi^*}{\to} A^* \overset{j^*}{\to} K^*$ is exact as well. Since the sequence $0 \to B^* \overset{\phi^*}{\to} A^* \to \operatorname{Im} j^* \to 0$ is exact, the desired result will follow from Theorem 12 if we show that $\operatorname{Im} j^*$ is semi-reflexive. But $\operatorname{Im} j^*$ is a submodule of K^* and K^* is semi-reflexive because it is a dual (Theorem 11). We may now deduce that $\operatorname{Im} j^*$ is semi-reflexive by appealing to Exercise 9.

Let $0 \to A_1 \to A \to A_2 \to 0$ be an exact sequence of left Λ-modules. This gives rise to a connecting homomorphism

$$\Delta : \operatorname{Hom}_\Lambda (A_1, \Lambda_l) \to \operatorname{Ext}^1_\Lambda (A_2, \Lambda_l).$$

In the first instance this is a homomorphism of abelian groups. However $\operatorname{Hom}_\Lambda (A_1, \Lambda_l)$ and $\operatorname{Ext}^1_\Lambda (A_2, \Lambda_l)$ are both of them right Λ-modules by virtue of the fact that Λ_l is a (Λ, Λ)-bimodule. Let λ belong to Λ. Then multiplication on the right by λ induces a Λ-homomorphism $\phi_\lambda : \Lambda_l \to \Lambda_l$ and, by (Chapter 3, Theorem 5), the diagram

$$
\begin{array}{ccc}
\operatorname{Hom}_\Lambda(A_1, \Lambda_l) & \overset{\Delta}{\longrightarrow} & \operatorname{Ext}^1_\Lambda(A_2, \Lambda_l) \\
\downarrow{\scriptstyle \operatorname{Hom}_\Lambda(A_1, \phi_\lambda)} & & \downarrow{\scriptstyle \operatorname{Ext}^1_\Lambda(A_2, \phi_\lambda)} \\
\operatorname{Hom}_\Lambda(A_1, \Lambda_l) & \overset{\Delta}{\longrightarrow} & \operatorname{Ext}^1_\Lambda(A_2, \Lambda_l)
\end{array}
$$

is commutative. Bearing in mind that $A_1^* = \mathrm{Hom}_\Lambda\,(A_1, \Lambda_l)$, this says that $\Delta\colon A_1^* \to \mathrm{Ext}_\Lambda^1\,(A_2, \Lambda)$ is a Λ-homomorphism and not just a homomorphism of abelian groups. Of course, we reach exactly the same conclusion if $0 \to A_1 \to A \to A_2 \to 0$ is an exact sequence in $\mathscr{C}_\Lambda^R$.

Theorem 14. *Let A be a finitely generated, semi-reflexive Λ-module. Then there exists a finitely generated, semi-reflexive Λ-module B (of the opposite type to A) with the aid of which exact sequences*

$$0 \to A \overset{\delta_A}{\to} A^{**} \to \mathrm{Ext}_\Lambda^1\,(B, \Lambda) \to 0 \qquad (5.4.1)$$

and
$$0 \to B \overset{\delta_A}{\to} B^{**} \to \mathrm{Ext}_\Lambda^1\,(A, \Lambda) \to 0 \qquad (5.4.2)$$

of Λ-modules, can be constructed.

Proof. We can form an exact sequence $0 \to K \to F \overset{\phi}{\to} A \to 0$ where F is a free module on a finite base. On applying the duality functor we obtain an exact sequence

$$0 \to A^* \overset{\phi^*}{\to} F^* \to B \to 0, \qquad (5.4.3)$$

where B is a certain Λ-module of the opposite type to A. But F^* is a free module on a finite base. Consequently B is finitely generated. Again, by Lemma 4, A^* is a closed submodule of F^*. Accordingly, by Theorem 12, B is semi-reflexive.

From (5.4.3) we now derive an exact sequence

$$0 \to B^* \to F^{**} \overset{\phi^{**}}{\to} A^{**} \to \mathrm{Ext}_\Lambda^1\,(B, \Lambda) \to 0 \qquad (5.4.4)$$

in $\mathscr{C}_\Lambda$ because, since F^* is free and hence projective, $\mathrm{Ext}_\Lambda^1\,(F^*, \Lambda) = 0$. Consider the commutative diagram

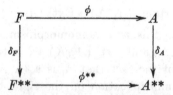

Here δ_F is an isomorphism and δ_A a monomorphism. It follows that A is isomorphic to $\delta_A(A) = \mathrm{Im}\,\phi^{**}$. Accordingly, by (5.4.4),

$$\mathrm{Coker}\,\delta_A = \mathrm{Ext}_\Lambda^1\,(B, \Lambda)$$

and we have an exact sequence

$$0 \to A \overset{\delta_A}{\to} A^{**} \to \mathrm{Ext}_\Lambda^1\,(B, \Lambda) \to 0. \qquad (5.4.5)$$

Next (5.4.4) yields an exact sequence $0 \to B^* \to F^{**} \to \operatorname{Im} \phi^{**} \to 0$ which we may replace by

$$0 \to B^* \to F^{**} \to A \to 0. \tag{5.4.6}$$

Put $M = \operatorname{Im} \phi^*$. Then (5.4.3) provides an exact sequence

$$0 \to M \to F^* \to B \to 0. \tag{5.4.7}$$

We can now repeat our argument but with $0 \to M \to F^* \to B \to 0$ replacing the original sequence $0 \to K \to F \to A \to 0$. If we do this, then (5.4.6) shows that we may use A to take over the role previously played by B. Hence in our new situation (5.4.5) becomes changed into an exact sequence

$$0 \to B \xrightarrow{\delta_B} B^{**} \to \operatorname{Ext}^1_\Lambda (A, \Lambda) \to 0$$

of Λ-modules. With this the proof is complete.

Our next theorem merely restates certain parts of Theorem 14 in a form which is particularly convenient for applications.

Theorem 15. *Given a finitely generated, semi-reflexive, Λ-module A, there exists a finitely generated, semi-reflexive, Λ-module B (of the opposite type to A) which can be used to form an exact sequence*

$$0 \to A \xrightarrow{\delta_A} A^{**} \to \operatorname{Ext}^1_\Lambda (B, \Lambda) \to 0$$

of Λ-modules. On the other hand given a finitely generated, semi-reflexive, Λ-module B, it is always possible to construct such an exact sequence with A finitely generated and semi-reflexive.

The reader should note that in order to obtain the second part of Theorem 15 it suffices to interchange the roles of A and B in Theorem 14.

5.5 Duality in Noetherian rings

In order to make further progress with our theory we shall, at this point, impose Noetherian conditions on Λ. We recall that if Λ is right Noetherian, then, by Theorem 5, every finitely generated right Λ-module is Noetherian and, by Exercise 7, the dual of any finitely generated left Λ-module is finitely generated. Of course, in this context, the roles of *left* and *right* are interchangeable.

Theorem 16. *Suppose that Λ is right Noetherian and that A is a finitely generated left Λ-module. Then A is semi-reflexive if and only if A is a submodule of a free module with a finite base.*

Proof. Suppose that A is semi-reflexive. Since Λ is right Noetherian, A^* is finitely generated. We can therefore construct an exact sequence $G \to A^* \to 0$, where G is a free Λ-module with a finite base. An application of the duality functor now yields an exact sequence $0 \to A^{**} \to G^*$. But, by hypothesis, $\delta_A : A \to A^{**}$ is a monomorphism. Accordingly A is a submodule of G^* (up to isomorphism) and G^* is a free module with a finite base. This proves half of the theorem. The other half follows from Exercise 9 and the fact that a free module with a finite base is reflexive.

Theorem 17. *Let Λ be left and right Noetherian. Then the following two statements are equivalent:*

(a) *Λ_l and Λ_r are both injective;*

(b) *all finitely generated left Λ-modules and all finitely generated right Λ-modules are reflexive.*

Proof. Assume (a). First suppose that A is finitely generated and semi-reflexive. Then, by Theorem 15, there exists an exact sequence

$$0 \to A \overset{\delta_A}{\to} A^{**} \to \mathrm{Ext}_\Lambda^1 (B, \Lambda) \to 0$$

of Λ-modules, for a suitable Λ-module B. It follows that δ_A is an isomorphism because, since Λ is injective, $\mathrm{Ext}_\Lambda^1 (B, \Lambda) = 0$. Thus finitely generated semi-reflexive modules are reflexive.

Now suppose that A is an *arbitrary* finitely generated Λ-module. It is then possible to construct an exact sequence $0 \to C \to F \to A \to 0$, where F is a free module with a finite base and C is a submodule of F. By Theorem 16, C is semi-reflexive. Further, since F is finitely generated it is Noetherian and therefore, by Theorem 3, C is finitely generated. Accordingly, by the remarks of the last paragraph, C is reflexive.

Consider the commutative diagram

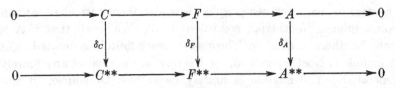

This has exact rows because the two dualizing functors are exact in the present instance, and δ_C and δ_F are isomorphisms. It follows that δ_A is an isomorphism and hence that A is reflexive. Thus (a) implies (b).

Assume (*b*). Let B be a finitely generated right Λ-module. Then, by hypothesis, B is reflexive and therefore *a fortiori* semi-reflexive. It follows, from Theorem 15, that there exists an exact sequence

$$0 \to A \overset{\delta_A}{\to} A^{**} \to \mathrm{Ext}^1_\Lambda (B, \Lambda_r) \to 0,$$

where A is finitely generated and semi-reflexive. However, by virtue of (*b*), δ_A is an isomorphism. Consequently $\mathrm{Ext}^1_\Lambda (B, \Lambda_r) = 0$. This holds for all finitely generated right Λ-modules B and, in particular, it holds whenever B is a cyclic right Λ-module. That Λ_r is injective now follows from (Chapter 3, Theorem 8). The proof that Λ_l is injective is similar.

Definition *The ring Λ is called a 'Quasi-Frobenius' ring if* (1) *it is left and right Noetherian, and* (2) *Λ_l and Λ_r are both injective.*

Thus if Λ is left and right Noetherian, then it is a Quasi-Frobenius ring if and only if all finitely generated Λ-modules are reflexive. It is known that if Λ is left and right Noetherian and Λ_l is injective, then Λ_r must be injective as well, but the usual proof of this depends on the structure theory of semi-simple rings and we shall not stop to go into this here. The reader will find details in J. P. Jans (10), Chapter 5.

The last theorem tells us when all finitely generated Λ-modules are reflexive. Next to be investigated will be the conditions under which all finitely generated, semi-reflexive modules are reflexive.

Theorem 18. *Let Λ be left and right Noetherian. Then the following two statements are equivalent*:

    (*a*) $r.\mathrm{Id}_\Lambda (\Lambda) \leqslant 1$;

    (*b*) *every finitely generated, semi-reflexive, left Λ-module is reflexive.*

Proof. Construct an exact sequence $0 \to \Lambda_r \to E \to U \to 0$ of right Λ-modules, where E is injective. Then $r.\mathrm{Id}_\Lambda (\Lambda) \leqslant 1$ if and only if U is injective.

Assume (*a*). Let A be a finitely generated, semi-reflexive, left Λ-module. By Theorem 15, there exists an exact sequence

$$0 \to A \overset{\delta_A}{\to} A^{**} \to \mathrm{Ext}^1_\Lambda (B, \Lambda_r) \to 0$$

of Λ-modules, where B is a finitely generated, semi-reflexive, right Λ-module. By Theorem 16, B is a submodule of a free module F with a finite base. Using the exact sequence $0 \to B \to F \to F/B \to 0$ and (Chapter 3, Theorem 13), we see that $\mathrm{Ext}^1_\Lambda (B, \Lambda_r)$ and $\mathrm{Ext}^1_\Lambda (F/B, U)$

are isomorphic. Since U is injective, this shows that $\operatorname{Ext}^1_\Lambda (B, \Lambda_r) = 0$ and hence that δ_A is an isomorphism. Our argument therefore shows that (a) implies (b).

Assume (b). Let B be a finitely generated right Λ-module. We can then construct an exact sequence $0 \to B_0 \to G \to B \to 0$ of right Λ-modules, where G is a free module with a finite base and B_0 is a submodule of G. Since B_0 is a submodule of a free module with a finite base it is semi-reflexive. It is also finitely generated because Λ is right Noetherian (see Theorem 3). It follows, from Theorem 15, that there exists an exact sequence

$$0 \to A \overset{\delta_A}{\to} A^{**} \to \operatorname{Ext}^1_\Lambda (B_0, \Lambda_r) \to 0,$$

where A is a finitely generated, semi-reflexive, left Λ-module. But such a module is reflexive by hypothesis. Consequently

$$\operatorname{Ext}^1_\Lambda (B_0, \Lambda_r) = 0.$$

However, by (Chapter 3, Theorem 13), $\operatorname{Ext}^1_\Lambda (B_0, \Lambda_r)$ and $\operatorname{Ext}^1_\Lambda (B, U)$ are isomorphic. Thus $\operatorname{Ext}^1_\Lambda (B, U) = 0$ and this holds for all finitely generated right Λ-modules B. From (Chapter 3, Theorem 8) we can now deduce that U is injective and hence that $r.\operatorname{Id}_\Lambda (\Lambda) \leqslant 1$. This completes the proof.

5.6 Perfect duality and Quasi-Frobenius rings

Suppose that Λ is left and right Noetherian. We have seen in Theorem 17 that in order that all finitely generated Λ-modules should be reflexive it is necessary and sufficient that Λ should be a Quasi-Frobenius ring. Such rings exhibit a rich duality theory of the kind encountered in the theory of finite-dimensional vector spaces. This situation may be described informally by the statement that for these rings we have *perfect duality* in respect of finitely generated modules. A summary of what this entails is provided by the next theorem. In it K denotes a submodule of a Λ-module A and M a submodule of the dual A^*. Moreover K^0 resp. M^0 denotes the annihilator of K resp. M in A^* resp. A. The conclusions of the theorem should be taken in conjunction with the more elementary results contained in Lemma 2.

Theorem 19. *Suppose that Λ is a Quasi-Frobenius ring. Further let A be a finitely generated Λ-module and A^* its dual (so A^* is also finitely generated). Then*

(1) $K = K^{00}$ for all submodules K of A;

(2) $0 \to K^0 \to A^* \to K^* \to 0$ is exact for every submodule K of A;

(3) $(K_1 \cap K_2)^0 = K_1^0 + K_2^0$, when K_1, K_2 are submodules of A;

(1*) $M = M^{00}$ for all submodules M of A;

(2*) $0 \to M^0 \to A \to M^* \to 0$ is exact for every submodule M of A;

(3*) $(M_1 \cap M_2)^0 = M_1^0 + M_2^0$, when M_1, M_2 are submodules of A^*.

Further if I is a left ideal of Λ and H a right ideal, then $I = I^{00}$ and $H = H^{00}$.

Remarks. In (2*) the homomorphism $A \to M^*$ is understood to be the one obtained by combining $A^{**} \to M^*$ (the dual of the inclusion mapping) with $\delta_A : A \to A^{**}$. We recall that I^{00} is the left annihilator of the right annihilator of I and H^{00} the right annihilator of the left annihilator of H.

Proof. For each submodule K of A we have an exact sequence $0 \to K \to A \to (A/K) \to 0$ and A/K is finitely generated and therefore reflexive. Accordingly, by Theorem 12, K is closed in A and therefore (1) is proved. Next, by Theorem 17, Λ_l and Λ_r are injective and therefore $0 \to (A/K)^* \to A^* \to K^* \to 0$ is exact. But, by Theorem 12, $(A/K)^*$ may be identified with K^0. Consequently (2) is proved as well. Further (1*) follows from (1) if we note that A is reflexive (because it is finitely generated) and make use of Lemma 3.

To establish (3) put $M_i = K_i^0$. Then, by Lemma 2 and (1),

$$(M_1 + M_2)^0 = M_1^0 \cap M_2^0 = K_1 \cap K_2$$

and therefore

$$(K_1 \cap K_2)^0 = (M_1 + M_2)^{00} = M_1 + M_2 = K_1^0 + K_2^0$$

by virtue of (1*). This proves (3) and the same device will serve to prove (3*).

Consider (2*). The exact sequence $0 \to M \to A^* \to (A^*/M) \to 0$ together with the injective character of Λ gives rise to the exact sequence

$$0 \to (A^*/M)^* \to A^{**} \to M^* \to 0.$$

Also, by Theorem 12, the image of $(A^*/M)^*$ in A^{**} is the annihilator of M in A^{**} and this, by Lemma 3, is $\delta_A(M^0)$ because A is reflexive. In view of the fact that $\delta_A : A \to A^{**}$ is an isomorphism, it follows that $0 \to M^0 \to A \to M^* \to 0$ is exact, where $A \to M^*$ is the homomorphism described in the remarks following the statement of the theorem. Accordingly (2*) has now been proved. This establishes the theorem because the assertions concerning I^{00} and H^{00} are special cases of (1).

Suppose momentarily that Λ is left and right Noetherian and that

$$I = I^{00}, \quad H = H^{00}$$

for all left ideals I and right ideals H. We shall see later that this is sufficient to ensure that Λ is a Quasi-Frobenius ring. However before we can win through to this result we must review briefly the notion of the *length* of a module.

Let C be a submodule of a Λ-module B and suppose that $C \neq B$. By a *composition series* from B to C we mean a chain

$$B = K_0 \supset K_1 \supset K_2 \supset \ldots \supset K_n = C \qquad (5.6.1)$$

of submodules from B to C with the property that each factor module $K_{i-1}/K_i \, (1 \leqslant i \leqslant n)$ is *simple*. If (5.6.1) is a composition series, then we say that it has *length* n. In the case where $C = B$ it is convenient to say that there is a composition series from B to C of length zero.

Exercise 12. *Let C be a submodule of a Λ-module B. Show that there exists a composition series from B to C if and only if B/C is both Noetherian and Artinian.*

Exercise 13. *Let C be a submodule of a Λ-module B and suppose that there exists a composition series from B to C of length p. Suppose also that*

$$B = K_0 \supset K_1 \supset \ldots \supset K_n = C$$

is a chain of submodules from B to C. Deduce that $n \leqslant p$. Hence show that any two composition series from B to C have the same length. Show also that any chain of submodules from B to C can be refined into a composition series (from B to C) by introducing extra terms into the chain.

Let B be a Λ-module. We put $L_\Lambda(B)$ equal to the length of a composition series from B to its zero submodule if such a series exists, otherwise we define $L_\Lambda(B)$ to be 'plus infinity'. Accordingly $L_\Lambda(B)$ is defined in all cases. It is called the *length* of B. Evidently isomorphic modules have the same length and there exists a composition series from B to a submodule C if and only if $L_\Lambda(B/C) < \infty$ in which case $L_\Lambda(B/C)$ will be the common length of all such composition series.†
By Exercise 12, $L_\Lambda(B) < \infty$ if and only if B is both Noetherian and

---

† This follows from the natural bijection connecting the submodules of B that contain C with the submodules of B/C.

Artinian. Note that $L_\Lambda(B) = 0$ if and only if $B = 0$ whereas $L_\Lambda(B) = 1$ if and only if B is a simple module.

Suppose that $L_\Lambda(B) < \infty$ and also $B \supset C \supset 0$, where C is a submodule of B. By Exercise 13, the chain $B \supset C \supset 0$ can be refined to a composition series

$$B = K_0 \supset K_1 \supset \ldots \supset K_q \supset K_{q-1} \supset \ldots \supset K_n = 0,$$

where $C = K_q$ say. Then $n = L_\Lambda(B)$, $n - q = L_\Lambda(C)$ and $q = L_\Lambda(B/C)$. Accordingly

$$L_\Lambda(B) = L_\Lambda(B/C) + L_\Lambda(C). \tag{5.6.2}$$

This holds provided that $L_\Lambda(B) < \infty$. Note that it is no longer necessary to exclude the possibility that C may coincide with B or its zero submodule for in either of these situations the assertion is trivial. Also (5.6.2) will still hold if B happens to be a zero module.

Now assume that $L_\Lambda(B) = \infty$ and that C is a submodule of B. Then either $L_\Lambda(B/C) = \infty$ or $L_\Lambda(C) = \infty$ for otherwise we could construct a composition series from B to C and one from C to 0 and these would combine to give a composition series from B to 0. Thus (5.6.2) is valid for *any* module B and submodule C and therefore we have proved

Theorem 20. *Let* $0 \to A' \to A \to A'' \to 0$ *be an exact sequence of* Λ-*modules. Then*

$$L_\Lambda(A) = L_\Lambda(A') + L_\Lambda(A'')$$

regardless of whether or not the lengths involved are finite.

Corollary. *Let* $A = A_1 \oplus A_2 \oplus \ldots \oplus A_n$ *in* $\mathscr{C}_A$. *Then*

$$L_\Lambda(A) = L_\Lambda(A_1) + L_\Lambda(A_2) + \ldots + L_\Lambda(A_n).$$

Proof. We need only consider the case $n = 2$ and this follows from the theorem because we have a canonical exact sequence

$$0 \to A_1 \to A_1 \oplus A_2 \to A_2 \to 0.$$

The next theorem shows that the notion of length is particularly relevant to our investigations relating to perfect duality.

Theorem 21. *Let* Λ *be a Quasi-Frobenius ring. If now A is a finitely generated Λ-module, then $L_\Lambda(A)$ is finite and, moreover, $L_\Lambda(A) = L_\Lambda(A^*)$, where A^* is the dual of A. In particular, $L_\Lambda(\Lambda_l)$ and $L_\Lambda(\Lambda_r)$ are finite and equal.*

Proof. By Theorem 19, there is a bijection between the submodules K of A and the submodules M of A^*. In this K and M correspond precisely when each is the annihilator of the other. Furthermore the bijection reverses inclusion relations.

Both A and $A*$ are finitely generated. Consequently, by Theorem 5, they are both Noetherian. Since $A*$ is Noetherian, the inclusion reversing bijection shows that A is Artinian as well as Noetherian. Accordingly $L_\Lambda(A) < \infty$. On replacing A by $A*$ we find that we also have $L_\Lambda(A*) < \infty$. Again, if we apply the bijection to a composition series from A to its zero submodule, the result will be a composition series from $A*$ to its zero submodule. This proves that $L_\Lambda(A) = L_\Lambda(A*)$ and now the proof is complete.

Theorem 22. *Suppose that $L_\Lambda(\Lambda_l)$ and $L_\Lambda(\Lambda_r)$ are both finite and that the dual of every simple module has length at most unity. Then Λ is a Quasi-Frobenius ring.*

Remark. It is in fact true, though we shall not stop to give a proof, that if Λ is left resp. right Artinian, then it is also left resp. right Noetherian.† It follows that $L_\Lambda(\Lambda_l)$ and $L_\Lambda(\Lambda_r)$ are both finite, if and only if Λ is left and right Artinian.

Proof. Our hypotheses ensure that Λ is both left Noetherian and left Artinian. They also ensure that it is right Noetherian and right Artinian. It follows, by Theorems 5 and 7, that every finitely generated Λ-module is Noetherian and Artinian and so has finite length.

Let A be a finitely generated Λ-module. We claim that

$$L_\Lambda(A*) \leqslant L_\Lambda(A).$$

This will be established by induction on $n = L_\Lambda(A)$. If $n = 0$ the assertion is trivial and if $n = 1$ it follows by hypothesis. We shall now suppose that $n > 1$ and make the obvious induction hypothesis. Let us construct an exact sequence $0 \to A_1 \to A \to A_2 \to 0$, where $L_\Lambda(A_1) = 1$ and $L_\Lambda(A_2) = n - 1$. On applying the duality functor we obtain an exact sequence

$$0 \to A_2^* \to A* \overset{\phi}{\to} A_1^*$$

of finitely generated Λ-modules and now, by Theorem 20 and the induction hypothesis,

$$
\begin{aligned}
L_\Lambda(A*) &= L_\Lambda(A_2^*) + L_\Lambda(\phi(A*)) \\
&\leqslant L_\Lambda(A_2^*) + L_\Lambda(A_1^*) \\
&\leqslant L_\Lambda(A_2) + L_\Lambda(A_1) \\
&= L_\Lambda(A).
\end{aligned}
$$

This establishes our claim.

† See, for example, (25) Theorem 3.25 Cor.

Since $\Lambda_l^* = \Lambda_r$ and $\Lambda_r^* = \Lambda_l$, it follows that $L_\Lambda(\Lambda_r) \leqslant L_\Lambda(\Lambda_l)$ and $L_\Lambda(\Lambda_l) \leqslant L_\Lambda(\Lambda_r)$. Hence $L_\Lambda(\Lambda_l) = L_\Lambda(\Lambda_r)$. Using Theorem 20 Cor., we now conclude that if F is a free module with a finite base, then

$$L_\Lambda(F) = L_\Lambda(F^*).$$

Once again let A be a finitely generated Λ-module. *This time we claim that $L_\Lambda(A^*) = L_\Lambda(A)$.* For we can construct an exact sequence $0 \to B \to F \to A \to 0$, where F is free with a finite base and B is finitely generated. Applying the duality functor we arrive at an exact sequence

$$0 \to A^* \to F^* \overset{\psi}{\to} B^*$$

whence
$$L_\Lambda(A^*) + L_\Lambda(\psi(F^*)) = L_\Lambda(F^*)$$
$$= L_\Lambda(F)$$
$$= L_\Lambda(A) + L_\Lambda(B).$$

But $L_\Lambda(A^*) \leqslant L_\Lambda(A)$ and $L_\Lambda(\psi(F^*)) \leqslant L_\Lambda(B^*) \leqslant L_\Lambda(B)$. Accordingly $L_\Lambda(A^*) = L_\Lambda(A)$ as required.

Let $0 \to A_1 \to A \to A_2 \to 0$ be an exact sequence of finitely generated Λ-modules. *We assert that the derived sequence $0 \to A_2^* \to A^* \to A_1^* \to 0$ is exact.* In any event we have an exact sequence

$$0 \to A_2^* \to A^* \overset{\omega}{\to} A_1^*$$

so our assertion will follow if we prove that ω is surjective. For this we need only show that $L_\Lambda(\omega(A^*)) = L_\Lambda(A_1^*)$. But

$$L_\Lambda(\omega(A^*)) = L_\Lambda(A^*) - L_\Lambda(A_2^*)$$
$$= L_\Lambda(A) - L_\Lambda(A_2)$$
$$= L_\Lambda(A_1)$$
$$= L_\Lambda(A_1^*),$$

which is what was needed.

Let I be a left ideal of Λ. If the main conclusion of the last paragraph is applied to the sequence $0 \to I \to \Lambda \to \Lambda/I \to 0$ (this is an exact sequence of finitely generated modules) we see in particular that the induced homomorphism $\operatorname{Hom}_\Lambda(\Lambda, \Lambda_l) \to \operatorname{Hom}_\Lambda(I, \Lambda_l)$ is surjective. It follows, by (Chapter 2, Theorem 11), that Λ_l is injective and a similar argument shows Λ_r to be injective as well. This proves the theorem.

Theorem 23. *Let Λ be left and right Noetherian. Then the following statements are equivalent:*

(a) for every left ideal I and right ideal H, $I^{00} = I$ and $H^{00} = H$;

(b) Λ is a Quasi-Frobenius ring.

Proof. *Assume* (a). There is then a bijection between the set of left ideals and the set of right ideals in which, when I corresponds to H, $I^0 = H$ and $H^0 = I$. This bijection reverses inclusion relations. Since Λ is right Noetherian, we may conclude that it must be left Artinian as well as left Noetherian. It follows that $L_\Lambda(\Lambda_l) < \infty$ and, of course, $L_\Lambda(\Lambda_r) < \infty$ for similar reasons.

Let A be a simple Λ-module. In view of Theorem 22, if we can show that the dual of A is also simple, then (b) will be proved. Clearly we need only consider the case where A is a left Λ-module and therefore we may suppose that $A = \Lambda/I$, where I is a maximal left ideal.

Consider the exact sequence $0 \to I \to \Lambda \to \Lambda/I \to 0$. By Theorem 12, $(\Lambda/I)^*$ is isomorphic to I^0 and our bijection shows that I^0 is a simple right ideal. Thus $(\Lambda/I)^*$ is a simple Λ-module and we have shown that (a) implies (b). The converse has already been established as a part of Theorem 19.

5.7 Group rings as Quasi-Frobenius rings

Some particularly interesting examples of Quasi-Frobenius rings arise in connection with the theory of finite groups. We shall now explain how this comes about.

Let G be a multiplicative group and let K be a field. By considering formal linear combinations of the elements of G we can construct, over K, a vector space $K(G)$ which has the elements of G as a base. Suppose that ξ belongs to $K(G)$. Then ξ has a unique representation in the form
$$\xi = \sum_{\sigma \in G} k_\sigma \sigma.$$

Here $k_\sigma \in K$ and only finitely many k_σ are non-zero. Let $\xi' = \sum_{\tau \in G} k'_\tau \tau$ be a second element of $K(G)$. We can turn $K(G)$ into a ring by defining the product of ξ and ξ' by means of the formula
$$\xi\xi' = \sum_{\sigma, \tau} k_\sigma k'_\tau (\sigma\tau).$$

This ring is the so-called *group ring* of G with coefficients in K. The identity element of the group ring is the identity element of G.

Let V be an arbitrary vector space over K. Since $K(G)$ may be considered as a $(K, K(G))$-bimodule with the elements of $K(G)$ operating on the right, $\mathrm{Hom}_K(K(G), V)$ has the structure of a left $K(G)$-module. (For similar reasons it may also be considered as a right $K(G)$-module.) Consequently if B is a left $K(G)$-module, then because it may be regarded as a K-space, it is possible to form two abelian groups namely $\mathrm{Hom}_K(B, V)$ and $\mathrm{Hom}_{K(G)}(B, \mathrm{Hom}_K(K(G), V))$. *We claim that there exists an isomorphism*

$$\mathrm{Hom}_K(B, V) \approx \mathrm{Hom}_{K(G)}(B, \mathrm{Hom}_K(K(G), V))$$

of abelian groups which is natural for $K(G)$-homomorphisms $B' \to B$.

The reader may remember that a very similar situation has been encountered earlier. Suppose that Λ is an arbitrary ring and that Θ is a module over the ring Z of integers. Then, as we saw in Example 3 of section (2.2), for each left Λ-module A there is an isomorphism

$$\mathrm{Hom}_Z(A, \Theta) \approx \mathrm{Hom}_\Lambda(A, \mathrm{Hom}_Z(\Lambda, \Theta))$$

of abelian groups which is natural for Λ-homomorphisms $A' \to A$. Indeed the arguments used to establish this result can be adapted in a straightforward way to meet the requirements of the new situation so we shall not go into any details.

It is now clear that $\mathrm{Hom}_K(-, V)$ and $\mathrm{Hom}_{K(G)}(-, \mathrm{Hom}_K(K(G), V))$, considered as functors from $\mathscr{C}^L_{K(G)}$ to $\mathscr{C}_Z$, are naturally equivalent. But all K-modules are injective† and therefore the functor $\mathrm{Hom}_K(-, V)$ is exact. Consequently our other functor is also exact and therefore $\mathrm{Hom}_K(K(G), V)$ *is an injective left $K(G)$-module.* For similar reasons $\mathrm{Hom}_K(K(G), V)$ is injective when regarded as a right $K(G)$-module.

Theorem 24. *Let K be a field and G a finite group. Then the group ring $K(G)$ is a Quasi-Frobenius ring.*

Proof. Since G has only a finite number of elements, the dimension of $K(G)$ as a K-space is finite. Let $I_1 \subseteq I_2 \subseteq I_3 \subseteq \ldots$ be an increasing sequence of left ideals of $K(G)$. Now each I_ν is a vector subspace of $K(G)$. It follows that the sequence terminates. Accordingly $K(G)$ is left Noetherian and similar considerations show it to be right Noetherian as well.

It will now suffice to show that $K(G)$ is injective both as a left $K(G)$-module and as a right $K(G)$-module. The first of these is typical so we shall concentrate on that. Now, considered as a left $K(G)$-module, $\mathrm{Hom}_K(K(G), K)$ is injective by our earlier remarks. If therefore we can show that $\mathrm{Hom}_K(K(G), K)$ is a free $K(G)$-module with a base consisting of a single element, this will show that

$$\mathrm{Hom}_K(K(G), K)$$

† See, for example, (Chapter 2, Exercise 11).

and $K(G)$ are isomorphic left $K(G)$-modules and the theorem will be established.

Let $\eta \in K(G)$. Then η is a linear combination, with coefficients in K, of the elements of G. Define a mapping $\mu: K(G) \to K$ by putting $\mu(\eta)$ equal to the coefficient of the identity element of G in the expression for η. Then $\mu \in \mathrm{Hom}_K(K(G), K)$.

Suppose that $f \in \mathrm{Hom}_K(K(G), K)$ and put $\xi = \sum\limits_{\tau \in G} f(\tau)\tau^{-1}$. Then $\xi \in K(G)$ and, for $\sigma \in G$,

$$(\xi \mu)(\sigma) = \mu(\sum\limits_{\tau \in G} \sigma f(\tau)\tau^{-1}) = f(\sigma).$$

Accordingly $\xi \mu = f$ and therefore $K(G)\mu = \mathrm{Hom}_K(K(G), K)$.

Now assume that ζ in $K(G)$ is such that $\zeta \mu = 0$. Let $\zeta = \sum\limits_{\sigma \in G} k_\sigma \sigma$. Then, for $\tau \in G, (\zeta \mu)(\tau^{-1}) = 0$ that is $\mu(\sum\limits_{\sigma \in G} \tau^{-1} k_\sigma \sigma) = 0$. Thus $k_\tau = 0$. It follows that $\zeta = 0$ and now we see that μ, by itself, constitutes a base for $\mathrm{Hom}_K(K(G), K)$ considered as a left $K(G)$-module. As already explained, the theorem follows.

Solutions to the Exercises on Chapter 5

Exercise 1. *Let Λ be left Noetherian. Show that every direct sum of injective left Λ-modules is injective.*

Solution. Let I be a left ideal of Λ and $f: I \to \bigoplus\limits_{s \in S} E_s$ a Λ-homomorphism, where $\{E_s\}_{s \in S}$ is a family of injective left Λ-modules. Since Λ is left Noetherian, I is finitely generated and therefore $f(I)$ is finitely generated as well. It follows that there exists a *finite* subset S_0, of S, such that $f(I) \subseteq \bigoplus\limits_{s \in S_0} E_s$. By restriction, we obtain a Λ-homomorphism $I \to \bigoplus\limits_{s \in S_0} E_s$, which, since $\bigoplus\limits_{s \in S_0} E_s$ is injective, can be extended to a Λ-homomorphism $\Lambda \to \bigoplus\limits_{s \in S_0} E_s$. It follows that f itself can be extended to a Λ-homomorphism $\Lambda \to \bigoplus\limits_{s \in S} E_s$. Since I was an arbitrary left ideal, it follows, from (Chapter 2, Theorem 11), that $\bigoplus\limits_{s \in S} E_s$ is an injective module.

Exercise 2. *Given that every countable direct sum of injective left Λ-modules is again injective, deduce that Λ is left Noetherian.*

Solution. Suppose that Λ is not left Noetherian. Then we can find a strictly ascending chain $I_1 \subset I_2 \subset I_3 \subset \ldots$ of left ideals of Λ. Put

$I = \bigcup\limits_{n=1}^{\infty} I_n$. Then I is also a left ideal of Λ and $I/I_n \neq 0$. For each n, let E_n be the injective envelope of $I/I_n, f_n : I/I_n \to E_n$ the inclusion mapping, and $\phi_n : I \to I/I_n$ the natural epimorphism. Then $g_n = f_n \phi_n$ is a non-zero Λ-homomorphism $g_n : I \to E_n$. Define a Λ-homomorphism $g : I \to \bigoplus\limits_{n=1}^{\infty} E_n$ by $g(\lambda) = \{g_n(\lambda)\}_{n=1}^{\infty}$. (This is well defined since, for each $\lambda \in I$, there exists n_0 such that, for $n \geqslant n_0, \lambda \in I_n$ and therefore

$$g_n(\lambda) = 0.)$$

By our hypothesis, $\bigoplus\limits_{n=1}^{\infty} E_n$ is injective and therefore g extends to a homomorphism $h : \Lambda \to \bigoplus\limits_{n=1}^{\infty} E_n$. Let $\pi_\nu : \bigoplus\limits_{n=1}^{\infty} E_n \to E_\nu$ be the canonical projection. Then there exists k such that $\pi_n h(1) = 0$ for all $n \geqslant k$. Suppose that $n \geqslant k$ and $\lambda \in I$. Then

$$g_n(\lambda) = \pi_n g(\lambda) = \pi_n h(\lambda) = \lambda \pi_n h(1) = 0.$$

Thus when $n \geqslant k$, $g_n : I \to E_n$ is a null homomorphism and we now have a contradiction. This completes the solution.

Exercise 3. *Suppose that A is a Noetherian left Λ-module and that x is an indeterminate. Show that the polynomial module $A[x]$ is Noetherian when considered as a left module with respect to the polynomial ring $\Lambda[x]$.*

Solution. Let U be a $\Lambda[x]$-submodule of $A[x]$. For $n = 0, 1, 2, \ldots$ let A_n consist of the zero element of A together with the leading coefficients of all polynomials of degree n that belong to U. Then A_n is a Λ-submodule of A and $A_n \subseteq A_{n+1}$ for all $n \geqslant 0$. Since A is a Noetherian Λ-module, we can choose an integer N such that $A_n = A_N$ whenever $n \geqslant N$. Now A_N is finitely generated. Let

$$A_N = \Lambda \alpha_1 + \Lambda \alpha_2 + \ldots + \Lambda \alpha_p$$

and for each i $(1 \leqslant i \leqslant p)$ select $\phi_i(x) \in U$ so that it has the form

$$\phi_i(x) = \alpha_i x^N + \beta_i x^{N-1} + \gamma_i x^{N-2} + \ldots.$$

Next $A + Ax + \ldots + Ax^{N-1}$, considered as a Λ-module, is isomorphic to the direct sum $A \oplus A \oplus \ldots \oplus A$, where there are N summands, and therefore it is Noetherian. It follows that

$$U \cap (A + Ax + \ldots + Ax^{N-1})$$

6

is a finitely generated Λ-module. Let it be generated by

$$\psi_1(x), \psi_2(x), \ldots, \psi_q(x).$$

Now suppose that $f(x) \in U$ and let it have degree k. Suppose for the moment that $k \geqslant N$. Then the leading coefficient of $f(x)$ belongs to $\Lambda \alpha_1 + \Lambda \alpha_2 + \ldots + \Lambda \alpha_p$. It follows that there exists $f_1(x) \in U$ such that $k_1 < k$, where k_1 is the degree of $f_1(x)$, and

$$f(x) \equiv f_1(x) \quad (\mathrm{mod}\ \Lambda[x]\,\phi_1 + \ldots + \Lambda[x]\,\phi_p).$$

If $k_1 \geqslant N$, then we can repeat the argument. Proceeding in this way we find that there exists $f^*(x) \in U$ such that the degree of $f^*(x)$ is smaller than N and $f(x) - f^*(x)$ belongs to $\Lambda[x]\,\phi_1(x) + \ldots + \Lambda[x]\,\phi_p(x)$. But then $f^*(x)$ is in $(A + Ax + \ldots + Ax^{N-1}) \cap U$. Consequently

$$f(x) \in \Lambda[x]\,\phi_1 + \ldots + \Lambda[x]\,\phi_p + \Lambda\psi_1 + \ldots + \Lambda\psi_q.$$

Accordingly U, considered as a $\Lambda[x]$-module, is generated by

$$\phi_1, \ldots, \phi_p, \psi_1, \ldots, \psi_q$$

and therefore, in particular, it is finitely generated.

Exercise 4. *Give an example of a ring which is left Noetherian but not right Noetherian.*

Solution. Let Λ be the ring generated, over the integers, by elements x, y which satisfy relations $yx = 0$ and $yy = 0$. Further let Γ be the subring $Z[x]$ of Λ. Then each element of Λ can be written uniquely in the form $\gamma_1 + \gamma_2 y$, where $\gamma_1, \gamma_2 \in \Gamma$.

By Exercise 3, Γ is a Noetherian ring. Now Λ is a finitely generated left Γ-module and therefore it is Noetherian as a left Γ-module. However if I is a left ideal of Λ, then I is a Γ-submodule of Λ and so it is finitely generated as a Γ-module. Consequently I is also finitely generated as a Λ-module. It follows that Λ is a left Noetherian ring.

Denote by I_n the right ideal of Λ generated by $y, xy, x^2y, \ldots, x^ny$. Then $\{I_n\}_{n \geqslant 1}$ is an increasing sequence of right ideals. Assume that there is an integer n such that $I_{n+1} = I_n$. Then

$$x^{n+1}y = y\lambda_0 + xy\lambda_1 + \ldots + x^ny\lambda_n,$$

where $\lambda_0, \lambda_1, \ldots, \lambda_n$ belong to Λ. But $yx = yy = 0$. Consequently for each $i = 0, 1, \ldots, n$ there is an integer k_i such that $y\lambda_i = yk_i = k_iy$. Thus $x^{n+1}y = k_0y + k_1xy + k_2x^2y + \ldots + k_nx^ny$. This however gives a contradiction because each element of Λ has a *unique* representation in the form $\gamma_1 + \gamma_2 y$ with $\gamma_1, \gamma_2 \in \Gamma$. Accordingly the I_n $(n \geqslant 1)$ form a strictly ascending chain of right ideals and therefore Λ cannot be right Noetherian.

Exercise 5. *Give an example of a ring which is left Artinian but not right Artinian.*

Solution.† Let K be a field and $\sigma: K \to K$ a ring-homomorphism such that $[K : \sigma(K)] = \infty$. (For example, K could consist of all rational functions over a field F in countably many indeterminates

$$x_1, x_2, x_3, \ldots$$

and $\sigma: K \to K$ could be defined so that it leaves each element of F fixed and satisfies $\sigma(x_\nu) = x_{\nu+1}$ for all ν.)

We now define a (K, K)-bimodule N with one K operating on the left and the other on the right. To this end let N have the same elements as K and the same addition. If now $n \in N$ and $k \in K$, then kn is to be obtained by using ordinary multiplication in K whereas nk is to be given by $nk = \sigma(k) n$. This produces a bimodule structure.

Put $\Lambda = K \oplus N$. Then Λ is an abelian group and it becomes a ring if we put
$$(k_1, n_1)(k_2, n_2) = (k_1 k_2, k_1 n_2 + n_1 k_2).$$

This has $(1, 0)$ as identity element and we can regard K as a subring of Λ if we identify k with $(k, 0)$. As a left K-space, Λ has dimension two. Consequently Λ is left Artinian.

Let $\{e_i\}_{i \in I}$ be a base for K over $\sigma(K)$ so that the index set I is infinite. Put $\nu_i = (0, e_i)$. Then $\nu_i \in N$. *We claim that, in Λ, the sum* $\sum_{i \in I} \nu_i \Lambda$ *of right ideals is direct.* This, of course, will show that Λ is neither right Artinian nor right Noetherian.

Suppose that $\sum \nu_i \lambda_i = 0$, where $\lambda_i = (k_i, n_i)$ and $\nu_i \lambda_i = 0$ for almost all i. Now $\nu_i \lambda_i = (0, \sigma(k_i) e_i)$ and therefore $\sigma(k_i) = 0$ for each i in I. Hence $k_i = 0$ and $\nu_i \lambda_i = 0$ in every case. This establishes our claim.

Exercise 6. *Show that if P is a projective Λ-module which can be generated by n elements, where $0 \leqslant n < \infty$, then the dual module P^* is also Λ-projective and it too can be generated by n elements.*

Solution. Let P be a projective module generated by $p_1, p_2, \ldots, p_n$ and let F be a free Λ-module with a base $e_1, e_2, \ldots, e_n$. There exists a Λ-epimorphism $\pi: F \to P$ with $\pi(e_i) = p_i$ for $1 \leqslant i \leqslant n$. Since P is

† This example is taken from A. Rosenberg and D. Zelinsky (23).

projective, there is a homomorphism $\sigma:P\to F$ such that $\pi\sigma = i_P$. Consequently $\sigma^*\pi^* = (\pi\sigma)^* = i_{P^*}$, where the notation is self-explanatory. It follows that P^* is a direct summand of F^* and therefore a homomorphic image of F^*. But, because F is a free module with a base of n elements, F^* is also a free module with a base of n elements. Accordingly P^* is a projective module which can be generated by n elements.

Exercise 7. *Let Λ be a right Noetherian ring. Show that the dual of any finitely generated left Λ-module is also finitely generated.*

Solution. Let A be a finitely generated left Λ-module, then we can construct an exact sequence $0\to K\to F\to A\to 0$, where F is a free left Λ-module with a finite base. On applying the duality functor we find that, up to isomorphism, A^* is a submodule of F^*. But F^* is a free right Λ-module with a finite base. Since Λ is right Noetherian, it follows, by Theorem 3, that A^* must be finitely generated.

Exercise 8. *Let R be an integral domain with quotient field F and $I \neq (0)$ an ideal of R. Show I is reflexive if and only if $I = (I^{-1})^{-1}$.*

Solution. The proof of Theorem 9 shows that there exists an isomorphism $\phi:I^* \xrightarrow{\sim} I^{-1}$ of R-modules where, if $f\in I^*$ and $\alpha\in I$ $(\alpha \neq 0)$, then $\phi(f) = f(\alpha)/\alpha$. It follows that $I^{**} \approx (I^{-1})^*$. Here ω in I^{**} corresponds to $\omega\phi^{-1}$ in $(I^{-1})^*$. By similar considerations we have an isomorphism $\psi:(I^{-1})^* \xrightarrow{\sim} (I^{-1})^{-1}$ of R-modules. This time if $g\in (I^{-1})^*$ and $\beta\in I^{-1}(\beta \neq 0)$, then $\psi(g) = g(\beta)/\beta$. Note that $I \subseteq (I^{-1})^{-1}$.

By combining ψ with the isomorphism $I^{**} \approx (I^{-1})^*$ we obtain an R-isomorphism $I^{**} \approx (I^{-1})^{-1}$. Note that if $\beta\in I^{-1}(\beta \neq 0)$, then ω in I^{**} will correspond to $\omega\phi^{-1}(\beta)/\beta$ in $(I^{-1})^{-1}$.

Consider the diagram

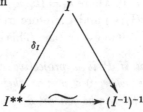

where $I\to (I^{-1})^{-1}$ is an inclusion mapping. Let $\alpha\in I, \beta\in I^{-1}$ with $\beta \neq 0$, and put $\omega = \delta_I(\alpha), \phi^{-1}(\beta) = f$. Then $f(\alpha) = \alpha\beta$ and therefore

$$\omega\phi^{-1}(\beta)/\beta = (\delta_I(\alpha))(f)/\beta = f(\alpha)/\beta = \alpha.$$

This proves that the diagram is commutative and now the desired result follows.

Exercise 9. *Show that a submodule of a semi-reflexive module is semi-reflexive and that a direct summand of a reflexive module is reflexive.*

Solution. Suppose that A is semi-reflexive and that B is a submodule of A. Let $\sigma: B \to A$ be the inclusion mapping. Then, with a self-explanatory notation,

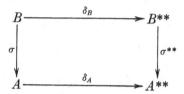

is a commutative diagram. But σ and δ_A are monomorphisms. Consequently δ_B is also a monomorphism and therefore B is semi-reflexive.

Now suppose that B is a direct summand of a reflexive module A. Then there exist homomorphisms $\sigma: B \to A$ and $\pi: A \to B$ such that $\pi\sigma = i_B$. It follows that $\pi^{**}\sigma^{**} = (\pi\sigma)^{**} = i_{B^{**}}$, and therefore π^{**} is an epimorphism. Next the diagram

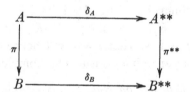

is commutative, and δ_A is an isomorphism. We now see that δ_B is an epimorphism. From the first part of this exercise δ_B is also a monomorphism. Accordingly δ_B is an isomorphism and B is reflexive.

Exercise 10. *Establish the relation $\delta_A^* \delta_{A^*} = i_{A^*}$.*

Solution. We have $\delta_{A^*}: A^* \to A^{***}$ and $\delta_A^*: A^{***} \to A^*$. Suppose that $f \in A^*$ and put $\phi = \delta_{A^*}(f)$. Then $\phi \in A^{***}$ and $\delta_A^*(\phi) = \phi\delta_A$. Thus $\delta_A^* \delta_{A^*}(f) = \delta_{A^*}(f)\delta_A$. Accordingly, for every a in A,

$$[\delta_A^* \delta_{A^*}(f)](a) = [\delta_{A^*}(f)\delta_A](a) = \delta_{A^*}(f)[\delta_A(a)].$$

But, by the definition of $\delta_{A^*}(f)$,

$$\delta_{A^*}(f)[\delta_A(a)] = [\delta_A(a)](f)$$
$$= f(a).$$

It follows that $\delta_A^* \delta_{A^*}(f) = f$ for all $f \in A^*$ and therefore $\delta_A^* \delta_{A^*} = i_{A^*}$.

Exercise 11. *Let $K_1, K_2, \ldots, K_n$ be submodules of a Λ-module A and suppose that A is their direct sum. Put*

$$U_i = K_1 + \ldots + K_{i-1} + K_{i+1} \ldots + K_n$$

so that U_i is a submodule of A. Show that $A^* = U_1^0 \oplus U_2^0 \oplus \ldots \oplus U_n^0$ and that U_i^0 and K_i^* are isomorphic Λ-modules.

Solution. Let $\sigma_i : K_i \to A$ and $\pi_i : A \to K_i$ be the canonical homomorphisms. Suppose that $f \in A^*$. Since $\pi_i(U_i) = 0$ we have $f\sigma_i \pi_i \in U_i^0$. Now $f = \sum\limits_{i=1}^{n} f\sigma_i \pi_i$ and therefore it follows that $A^* = U_1^0 + U_2^0 + \ldots + U_n^0$.

Assume next that for $i = 1, 2, \ldots, n$, $f_i \in U_i^0$ and $f_1 + f_2 + \ldots + f_n = 0$. For any $a \in A$ we have $a = k_1 + k_2 + \ldots + k_n$ for suitable $k_i \in K_i$, and, since $K_i \subseteq U_j$ whenever $i \neq j$, $f_j(k_i) = 0$ provided that $i \neq j$. Thus $f_j(a) = f_j(k_1) + f_j(k_2) + \ldots + f_j(k_n) = f_j(k_j)$ and therefore

$$-f_j(a) = f_1(k_j) + \ldots + f_{j-1}(k_j) + f_{j+1}(k_j) + \ldots + f_n(k_j) = 0.$$

This shows that $f_j = 0$ for every j. Consequently in the relation $A^* = U_1^0 + U_2^0 + \ldots + U_n^0$ the sum is direct.

In order to show that U_i^0 and K_i^* are isomorphic define a Λ-homomorphism $\phi : U_i^0 \to K_i^*$ by $\phi(f) = f\sigma_i$. If now $g \in K_i^*$, then

$$g\pi_i \in U_i^0 \quad \text{and} \quad \phi(g\pi_i) = g\pi_i \sigma_i = g.$$

Hence ϕ is an epimorphism. Finally if $f \in U_i^0$ and $\phi(f) = 0$, then f vanishes not only on $K_1, \ldots, K_{i-1}, K_{i+1}, \ldots, K_n$ but also on K_i because $f\sigma_i = \phi(f) = 0$. It follows that $f = 0$ and hence that ϕ is a monomorphism as well as an epimorphism. This completes the solution.

Exercise 12. *Let C be a submodule of a Λ-module B. Show that there exists a composition series from B to C if and only if B/C is both Noetherian and Artinian.*

Solution. Suppose that $B = K_0 \supset K_1 \supset \ldots \supset K_n = C$ is a composition series from B to C. If $n = 1$, then B/C is a simple module and therefore Noetherian and Artinian. Suppose now that a similar implication has ben established for any composition series of length $n - 1$. Then, since $B = K_0 \supset K_1 \supset \ldots \supset K_{n-1}$ is a composition series from B to K_{n-1}, we see that B/K_{n-1} is Noetherian and Artinian. Also K_{n-1}/C is simple and hence Noetherian and Artinian. Since we have an exact sequence $0 \to K_{n-1}/C \to B/C \to B/K_{n-1} \to 0$ the desired conclusion, namely that B/C is Noetherian and Artinian, follows.

Now assume that B/C is Noetherian and Artinian. Let S be the set of submodules K of B such that (1) $B \supseteq K \supseteq C$, and (2) there exists a composition series from K to C. Since C belongs to S, S is not empty. By considering the submodules K/C of B/C we see that S contains a maximal member B' say. It is enough to show that $B' = B$. Assume that $B' \ne B$. Then B/B' is non-zero and it is Artinian because it is isomorphic to $(B/C)/(B'/C)$. It follows that there exists a submodule B'' of B such that $B \supseteq B'' \supset B'$ and there are no submodules of B strictly between B'' and B'. Accordingly $B'' \in S$ and now we have the desired contradiction since B' is maximal in S.

Exercise 13. *Let C be a submodule of a Λ-module B, and suppose that there exists a composition series from B to C of length p. Suppose also that $B = K_0 \supset K_1 \supset \ldots \supset K_n = C$ is any chain of submodules from B to C. Deduce that $n \leqslant p$. Hence show that any two composition series from B to C have the same length. Show also that any chain of submodules from B to C can be refined into a composition series (from B to C) by introducing extra terms into the chain.*

Solution. We prove the first assertion by induction on p. Clearly if $p = 0$ the result is obvious as in that case $B = C$. Suppose that $p > 0$ and let $B = J_0 \supset J_1 \supset \ldots \supset J_p = C$ be a composition series from B to C. Clearly we may suppose that $n > 0$. Consider the modules

$$B = J_0 + K_{n-1} \supseteq J_1 + K_{n-1} \supseteq \ldots \supseteq J_p + K_{n-1} = K_{n-1}.$$

Since we have isomorphisms

$$(J_i + K_{n-1})/(J_{i+1} + K_{n-1}) \approx ((J_{i+1} + K_{n-1}) + J_i)/(J_{i+1} + K_{n-1})$$
$$\approx J_i/(J_i \cap (J_{i+1} + K_{n-1}))$$
$$\approx J_i/(J_{i+1} + J_i \cap K_{n-1})$$
$$\approx (J_i/J_{i+1})/((J_{i+1} + J_i \cap K_{n-1})/J_{i+1})$$

we see that $(J_i + K_{n-1})/(J_{i+1} + K_{n-1})$ is either a simple module or a zero module; furthermore if $J_i \cap K_{n-1} \nsubseteq J_{i+1}$ then it is a zero module. *We claim that there is at least one value of i for which*

$$J_i \cap K_{n-1} \nsubseteq J_{i+1}.$$

For we can choose $b \in K_{n-1}$ so that $b \notin K_n$. Now select i as large as possible so that $b \in J_i$. Then $i < p$. Also $b \in J_i \cap K_{n-1}$ but $b \notin J_{i+1}$. This establishes our claim.

Our discussion so far shows that there must exist a composition series from B to K_{n-1} of length p', where $p' < p$. By induction we conclude that $n - 1 \leqslant p' \leqslant p - 1$. Accordingly $n \leqslant p$ and the first assertion is proved. The other assertions are immediate consequences.

6

LOCAL HOMOLOGICAL ALGEBRA

6.1 Notation

As usual, Λ denotes a ring with an identity element and we do not assume that Λ is commutative. When we wish to consider Λ as a left respectively right Λ-module, we denote it by Λ_l respectively Λ_r. In the latter part of this chapter, it will be necessary to pay special attention to the properties of *commutative* rings. A typical commutative ring (with identity element) will be denoted by S.

6.2 Projective covers

In this section we shall introduce the concept of a *projective cover*. This is the counterpart of the notion of an *injective envelope* which was defined in section (2.6). However, whereas a Λ-module always possesses an injective envelope, it is only in special circumstances that a projective cover will exist.†

Let B be a Λ-module.

Definition. *A 'projective cover' for B is a pair (P, ψ), where* (i) *P is a projective Λ-module,* (ii) *$\psi : P \to B$ is a Λ-epimorphism, and* (iii) *no proper submodule of P is mapped by ψ on to B.*

For example, a projective module is its own projective cover with respect to the identity mapping. It should be noted that in some situations we omit a direct reference to the epimorphism ψ, but when this happens it will be clear from the context which is the relevant mapping.

The next theorem shows that projective covers (when they exist) are essentially unique.

Theorem 1. *Let $u : B \xrightarrow{\sim} B'$ be an isomorphism of Λ-modules and let (P, ψ) resp. (P', ψ') be a projective cover for B resp. B'. Then there exists an isomorphism $v : P \xrightarrow{\sim} P'$, of Λ-modules, which satisfies $u\psi = \psi'v$.*

† An account of rings for which every module has a projective cover will be found in H. Bass (3).

Proof. Since P is projective and ψ' is an epimorphism, we can find a Λ-homomorphism $v: P \to P'$ such that $\psi'v = u\psi$. Then

$$\psi'v(P) = u\psi(P) = u(B) = B'$$

and therefore $v(P) = P'$, because (P', ψ') is a projective cover for B'. Since P' is projective and we have just shown that v is an epimorphism, we must have $P = \operatorname{Ker} v \oplus C$ for a suitable submodule C of P. It follows that

$$u\psi(C) = \psi'v(C) = \psi'v(P) = \psi'(P') = B' = u(B)$$

whence $\psi(C) = B$ because u is an isomorphism. Since (P, ψ) is a projective cover for P, this shows that $C = P$ and therefore $\operatorname{Ker} v = 0$. We have now proved that v is an isomorphism, and with this the theorem is established.

Let B be a Λ-module and

$$\ldots \to P_n \to P_{n-1} \to \ldots \to P_1 \to P_0 \to B \to 0 \qquad (6.2.1)$$

a projective resolution of B. Put $B_0 = B$ and, for $n \geqslant 1$, put

$$B_n = \operatorname{Im}(P_n \to P_{n-1}).$$

Definition. *The exact sequence* (6.2.1) *is called a 'minimal projective resolution' of B if, for each $i \geqslant 0$, P_i is a projective cover for B_i.*

Theorem 2. *Suppose that*

$$\ldots \to P_n \to P_{n-1} \to \ldots \to P_0 \to B \to 0$$

is a minimal projective resolution of the Λ-module B, and let $m \geqslant 0$ be an integer. Then, with the usual notation for projective dimension, $\operatorname{Pd}_\Lambda(B) < m$ if and only if $P_n = 0$ for all $n \geqslant m$.

This is nearly obvious. Since a closely analogous result is to be found in (Chapter 3, Theorem 20), we shall not go into details.

Minimal projective resolutions have strong uniqueness properties. This is shown by

Theorem 3. *Let $u: B \xrightarrow{\sim} B'$ be an isomorphism of Λ-modules and let*

$$\ldots \to P_n \to \ldots \to P_1 \to P_0 \to B \to 0$$

respectively $\qquad \ldots \to P_n' \to \ldots \to P_1' \to P_0' \to B' \to 0$

be a minimal projective resolution of B respectively B'. Then there exist

Λ-*isomorphisms* $\psi_i : P_i \xrightarrow{\sim} P'_i$ $(i = 0, 1, 2, \ldots)$ *which make*

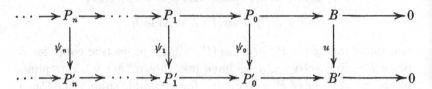

a commutative diagram

This is a simple application† of Theorem 1.

6.3 Quasi-local and local rings

We recall that an element u, of Λ, is called a *unit* if there exists $v \in \Lambda$ such that $uv = vu = 1$. Of course, v (when it exists) is unique and is the *inverse* of u.

Definition. *A ring Λ is called a 'quasi-local ring' if the non-units of Λ form a two-sided ideal.*

Suppose that Λ is a quasi-local ring and that J is the ideal formed by the non-units. Evidently J is a maximal left ideal and, indeed, Λ has no other maximal left ideal. Likewise, J is also the unique maximal right ideal of Λ. Now, in any ring, the intersection of all the maximal left ideals coincides with the intersection of all the maximal right ideals. The common intersection is known as the *Jacobson radical* of the ring. In the case of our quasi-local ring, it is clear that the Jacobson radical is simply the ideal J of non-units. For this reason the ideal of non-units in a quasi-local ring will be referred to as the *radical* of the ring.

Let Λ be a quasi-local ring and J its radical. If I is a two-sided ideal of Λ and $I \neq \Lambda$, then $I \subseteq J$ and Λ/I is a quasi-local ring with radical J/I. In particular Λ/J has the zero ideal as its radical and therefore every non-zero element of Λ/J is a unit in that ring. In other terms, Λ/J *is a division ring*. Any element in Λ which is not in J is a unit. Accordingly if $c \in J$, then $1 - c$ is a unit of Λ.

Theorem 4.‡ *Let Λ be a quasi-local ring with radical J and let M be a finitely generated left Λ-module. Assume that K is a submodule of M such that $K + JM = M$. Then $K = M$. In particular if $JM = M$, then $M = 0$.*

† Compare with (Chapter 3, Theorem 19).
‡ This result is a form of *Nakayama's Lemma*.

Proof. We have $J(M/K) = M/K$ and M/K is finitely generated. Moreover it is enough to show that $M/K = 0$. Accordingly we may suppose that $K = 0$. Thus we assume that $JM = M$ and deduce that M is a null module.

Suppose that $M \neq 0$, and let $M = \Lambda u_1 + \ldots + \Lambda u_{n-1} + \Lambda u_n$ with n minimal. Then u_n belongs to $M = JM$ so

$$u_n = c_1 u_1 + \ldots + c_{n-1} u_{n-1} + c_n u_n$$

with $c_i \in J$. We now have

$$(1 - c_n) u_n \in \Lambda u_1 + \Lambda u_2 + \ldots + \Lambda u_{n-1}$$

and therefore $u_n \in \Lambda u_1 + \Lambda u_2 + \ldots + \Lambda u_{n-1}$ because $1 - c_n$ is a unit. It follows that $M = \Lambda u_1 + \Lambda u_2 + \ldots + \Lambda u_{n-1}$ and this contradicts the minimal property of n. The proof is thus complete.

Let M be a Λ-module and $u_1, u_2, \ldots, u_n$ elements of M. We shall say that $u_1, u_2, \ldots, u_n$ form a *minimal set of generators* for M if (i) $u_1, u_2, \ldots, u_n$ generate M, and (ii) for no value of i $(1 \leqslant i \leqslant n)$ is M generated by $u_1, \ldots, u_{i-1}, u_{i+1}, \ldots, u_n$.

Exercise 1. *Let Λ be a quasi-local ring with radical J, M a finitely generated left Λ-module, and $u_1, u_2, \ldots, u_n$ elements of M. Denote by $\bar{u}_i$ the natural image of u_i in M/JM. Show that $u_1, u_2, \ldots, u_n$ generate M if and only if $\bar{u}_1, \bar{u}_2, \ldots, \bar{u}_n$ generate M/JM over the division ring Λ/J. Show also that $u_1, u_2, \ldots, u_n$ is a minimal set of generators for M if and only if $\bar{u}_1, \bar{u}_2, \ldots, \bar{u}_n$ is a base for M/JM over Λ/J. Deduce that any two minimal sets of generators for M contain the same number of elements, and that this number is equal to the length of M/JM considered either as a Λ-module or as a Λ/J-module.*

A useful application of Exercise 1 is provided by

Exercise 2. *Λ is a quasi-local ring with radical J and M is a finitely generated left Λ-module. Furthermore $u \in M$ but $u \notin JM$. Show that there exists a minimal set of generators for M that has u as a member.*

We are now ready to consider the question of the existence of projective covers in the quasi-local case.

Lemma 1. *Let Λ be a quasi-local ring with radical J and let $\psi : P \to M$ be an epimorphism, where P is a finitely generated, projective, left Λ-module. If now $\mathrm{Ker}\, \psi \subseteq JP$, then (P, ψ) is a projective cover for M.*

Proof. Let X be a submodule of P such that $\psi(X) = M$. Then $X + \mathrm{Ker}\, \psi = P$ and therefore $X + JP = P$. Consequently, by Theorem 4, $X = P$ and now the proof is complete.

Theorem 5. *Let Λ be a quasi-local ring with radical J and M a finitely generated left Λ-module. Then M has a projective cover. If (P, ψ) is a projective cover for M, then P is a free module on a finite base and $\operatorname{Ker} \psi \subseteq JP$.*

Proof. Let $M = \Lambda u_1 + \Lambda u_2 + \ldots + \Lambda u_n$, where the integer n is minimal, and construct a free module F with a base $e_1, e_2, \ldots, e_n$ of n elements. Define an epimorphism $\psi : F \to M$ so that $\psi(e_i) = u_i$ for $1 \leqslant i \leqslant n$. We claim that $\operatorname{Ker} \psi \subseteq JF$. For suppose that

$$\lambda_1 e_1 + \lambda_2 e_2 + \ldots + \lambda_n e_n \in \operatorname{Ker} \psi.$$

We must show that all the λ_i are in J. Assume the contrary. Without loss of generality we may suppose that $\lambda_1 \notin J$, i.e. we may suppose that λ_1 is a unit. Since $\lambda_1 u_1 + \lambda_2 u_2 + \ldots + \lambda_n u_n = 0$, it follows that $u_1 \in \Lambda u_2 + \ldots + \Lambda u_n$. Accordingly $M = \Lambda u_2 + \ldots + \Lambda u_n$ and, as this contradicts the minimality of n, our claim is established.

Lemma 1 now shows that (F, ψ) is a projective cover for M. This particular projective cover has the properties described in the statement of the theorem. The uniqueness of projective covers shows that all other projective covers share these properties.

Theorem 6. *Let Λ be a quasi-local ring and P a finitely generated projective Λ-module. Then P is free.*

Proof. The pair (P, i_P) is a projective cover for P. Hence, by Theorem 5, P is a free Λ-module.

It is known that any projective module whatsoever over a quasi-local ring is free, but the proof of this involves considerations of a different kind, and we shall not go into details here.†

Exercise 3. *Let I be a two-sided ideal of an arbitrary ring Λ and let (P, ψ) be a projective cover for a left Λ-module A. Show that $(P/IP, \psi^*)$ is a projective cover for the Λ/I-module A/IA, where $\psi^* : P/IP \to A/IA$ is the homomorphism induced by ψ. (Consequently if γ belongs to the centre of Λ, then $P/\gamma P$ is a projective cover for the $\Lambda/\gamma\Lambda$-module $A/\gamma A$.)*

This is a convenient point at which to introduce Noetherian conditions.

Definition. *A ring Λ is called a 'left local ring' if it is quasi-local and left Noetherian.*

Of course, if Λ is a quasi-local and right Noetherian, then it is called a *right* local ring. We give an example.

† For a proof see I. Kaplansky (11).

Let $x_1, x_2, ..., x_s$ be indeterminates and denote by $\Lambda[[x_1, x_2, ..., x_s]]$ the ring of *formal power series* in $x_1, x_2, ..., x_s$ with coefficients in Λ. Addition and multiplication of formal power series are defined in the obvious way, it being understood that the indeterminates commute with each other and with the elements of Λ.

Exercise 4. *Show that a member of $\Lambda[[x_1, x_2, ..., x_s]]$ is a unit of that ring if and only if its constant term is a unit of Λ. Show also that if Λ is left Noetherian, then so too is $\Lambda[[x_1, x_2, ..., x_s]]$.*

It follows, from this exercise, that if Λ is a left local ring, then $\Lambda[[x_1, x_2, ..., x_s]]$ is a left local ring as well. More particularly, if D is a division ring, then $D[[x_1, x_2, ..., x_s]]$ is a left and right local ring and the non-units are just the power series which have zero constant terms.

Let Λ be a left local ring and M a finitely generated left Λ-module. By Theorem 5, M has a projective cover (F_0, ψ_0), where F_0 is a free module on a finite base. Put $M_1 = \text{Ker}\,\psi_0$. Since Λ is left Noetherian, it follows, by (Chapter 5, Theorem 5 Cor.), that M_1 is finitely generated. Accordingly it has a projective cover (F_1, ψ_1), where F_1 is a free module with a finite base. Put $M_2 = \text{Ker}\,\psi_1$. Then M_2 is finitely generated and it has a projective cover (F_2, ψ_2). Evidently we can continue in this way indefinitely. Define $d_i : F_i \to F_{i-1}$ to be $\psi_i : F_i \to M_i$ followed by the inclusion mapping $M_i \to F_{i-1}$. Then

$$... \to F_i \overset{d_i}{\to} F_{i-1} \to ... \to F_1 \overset{d_1}{\to} F_0 \overset{\psi_0}{\to} M \to 0$$

is a minimal projective resolution of M. Thus every finitely generated left Λ-module has a minimal projective resolution and each projective module appearing in such a resolution is a free module on a finite base.

Theorem 7. *Let Λ be a left local ring, M a finitely generated left Λ-module, and*

$$... \to P_n \to P_{n-1} \to ... \to P_1 \to P_0 \to M \to 0$$

a minimal projective resolution of M. Further let γ be a central element in the radical J (of Λ) which is not a zerodivisor on either Λ or M. Then the derived sequence

$$... \to P_n/\gamma P_n \to P_{n-1}/\gamma P_{n-1} \to ... \to P_1/\gamma P_1 \to P_0/\gamma P_0 \to M/\gamma M \to 0$$

is a minimal projective resolution of the $\Lambda/\gamma\Lambda$-module $M/\gamma M$.

Proof. Put $M_0 = M$ and, for $i \geqslant 1$, put $M_i = \text{Im}\,(P_i \to P_{i-1})$. Then for each $i \geqslant 1$ we have a commutative diagram

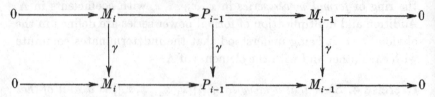

with exact rows, where each vertical mapping is produced by multiplication by γ. Since γ is not a zerodivisor on Λ, it is not a zerodivisor on any submodule of a free Λ-module. In particular γ is not a zerodivisor on P_i $(i \geqslant 0)$ nor on any M_j provided that $j \geqslant 1$. Moreover, by hypothesis, γ is also not a zerodivisor on $M_0 = M$. It follows that each vertical mapping in our diagram is a monomorphism and therefore, applying the theory of the Ker–Coker sequence,

$$0 \to M_i/\gamma M_i \to P_{i-1}/\gamma P_{i-1} \to M_{i-1}/\gamma M_{i-1} \to 0$$

is exact for every i. Accordingly the infinite sequence

$$\ldots \to P_1/\gamma P_1 \to P_0/\gamma P_0 \to M/\gamma M \to 0$$

is exact. Finally, by Exercise 3, $P_i/\gamma P_i$ is a projective cover for $M_i/\gamma M_i$ and with this the proof is complete.

Theorem 8. *Let Λ be a left local ring with radical J, and M a finitely generated left Λ-module. Further let γ be an element of the centre of Λ which is contained in J, and suppose that γ is not a zerodivisor on Λ or M. In these circumstances $l.\mathrm{Pd}_\Lambda(M) = l.\mathrm{Pd}_{\Lambda/\gamma\Lambda}(M/\gamma M)$.*

Proof. Let the notation be as in Theorem 7. By Theorem 4, $M = 0$ if and only if $M/\gamma M = 0$ and, by the same result, $P_n = 0$ if and only if $P_n/\gamma P_n = 0$. Theorem 8 now follows by virtue of Theorems 2 and 7.

The next theorem shows that when the radical of a local ring has a non-zero annihilator, all finitely generated modules having finite projective dimension are free.

Theorem 9. *Let Λ be a left local ring with radical J, and M a finitely generated left Λ-module such that $l.\mathrm{Pd}_\Lambda(M) < \infty$. If now there exists $\lambda \in \Lambda$ such that $\lambda \neq 0$ and $\lambda J = 0$, then M is a free module.*

Proof. Put $l.\mathrm{Pd}_\Lambda(M) = n$. By Theorem 6, it is enough to show that M is projective, i.e. that $n \leqslant 0$.

Assume that $n \geqslant 1$. We now define a left Λ-module K. If $n = 1$ we put $K = M$. If however $n \geqslant 2$, then we construct an exact sequence

$$0 \to K \to P_{n-2} \to \ldots \to P_1 \to P_0 \to M \to 0,$$

where each P_i is finitely generated and projective. In either case K is finitely generated and $l.\mathrm{Pd}_\Lambda(K) = 1$. The theorem will follow if we obtain a contradiction.

Let (F, ψ) be a projective cover for K. Then the sequence

$$0 \to \mathrm{Ker}\,\psi \to F \xrightarrow{\psi} K \to 0$$

is exact and, since $l.\mathrm{Pd}_\Lambda(K) = 1$, $\mathrm{Ker}\,\psi$ is projective and therefore, by Theorem 6, free. Again, by Theorem 5, $\mathrm{Ker}\,\psi \subseteq JF$. Consequently $\lambda \,\mathrm{Ker}\,\psi = 0$ because $\lambda J = 0$. It now follows, because $\lambda \neq 0$ and $\mathrm{Ker}\,\psi$ is free, that $\mathrm{Ker}\,\psi = 0$. Accordingly F and K are isomorphic and now we have the required contradiction because $l.\mathrm{Pd}_\Lambda(K) = 1$.

Let Λ be a local ring with radical J. Then Λ/J is a Λ-module and it is not unreasonable to expect that it will have special importance in relation to the homological properties of Λ. We shall now embark on some investigations which lend support to this idea and which will culminate in a demonstration that the global dimension of Λ is equal to the injective dimension of Λ/J.

Lemma 2. *Let Λ be a quasi-local ring with radical J, and suppose that C is a finitely generated left Λ-module such that $\mathrm{Hom}_\Lambda(C, \Lambda/J) = 0$. Then $C = 0$.*

Proof. Let us assume that $C \neq 0$. By Theorem 4, $C/JC \neq 0$ and we have an exact sequence

$$0 \to \mathrm{Hom}_\Lambda(C/JC, \Lambda/J) \to \mathrm{Hom}_\Lambda(C, \Lambda/J).$$

The lemma will therefore follow if we show that $\mathrm{Hom}_\Lambda(C/JC, \Lambda/J) \neq 0$. But C/JC is a free Λ/J-module with a base $e_1, e_2, \ldots, e_n$, where $n \geqslant 1$. Hence C/JC is isomorphic to $\Lambda/J \oplus \Lambda/J \oplus \ldots \oplus \Lambda/J$, where there are n summands, and therefore we have an isomorphism

$$\mathrm{Hom}_\Lambda(C/JC, \Lambda/J) \approx \bigoplus_{i=1}^{n} \mathrm{Hom}_\Lambda(\Lambda/J, \Lambda/J).$$

Since $\mathrm{Hom}_\Lambda(\Lambda/, \Lambda/J) \neq 0$, the lemma follows.

Theorem 10. *Let Λ be a left local ring with radical J and C a finitely generated left Λ-module. Then C is projective if and only if*

$$\mathrm{Ext}^1_\Lambda(C, \Lambda/J) = 0.$$

Proof. We shall assume that $\mathrm{Ext}^1_\Lambda(C, \Lambda/J) = 0$ and deduce that C is projective. The converse follows by virtue of (Chapter 3, Theorem 1).

Let (F, ψ) be a projective cover for C and put $K = \mathrm{Ker}\, \psi$. Then

$$0 \to K \to F \overset{\psi}{\to} C \to 0 \qquad\qquad (6.3.1)$$

is an exact sequence and, by Theorem 5, $K \subseteq JF$ and F is a free module on a finite base. Of course, since Λ is left Noetherian, K is finitely generated.

By hypothesis, $\mathrm{Ext}_\Lambda^1 (C, \Lambda/J) = 0$ and therefore the sequence

$$0 \to \mathrm{Hom}_\Lambda (C, \Lambda/J) \to \mathrm{Hom}_\Lambda (F, \Lambda/J) \to \mathrm{Hom}_\Lambda (K, \Lambda/J) \to 0,$$

to which (6.3.1) gives rise, is exact. But $K \subseteq JF \subseteq F$. Consequently the mapping $\mathrm{Hom}_\Lambda (F, \Lambda/J) \to \mathrm{Hom}_\Lambda (K, \Lambda/J)$ can be factored into

$$\mathrm{Hom}_\Lambda (F, \Lambda/J) \overset{\xi}{\to} \mathrm{Hom}_\Lambda (JF, \Lambda/J) \overset{\eta}{\to} \mathrm{Hom}_\Lambda (K, \Lambda/J).$$

However if $f \in \mathrm{Hom}_\Lambda (F, \Lambda/J)$, then $\xi(f)$, which is the restriction of f to JF, is null because J annihilates Λ/J. Thus ξ is a null mapping and therefore the *epimorphism* $\eta\xi: \mathrm{Hom}_\Lambda (F, \Lambda/J) \to \mathrm{Hom}_\Lambda (K, \Lambda/J)$ is null as well. It follows that $\mathrm{Hom}_\Lambda (K, \Lambda/J) = 0$ and therefore $K = 0$ by Lemma 2. This shows that $\psi: F \to C$ is an isomorphism and proves that C is free as required.

Let Λ be a left local ring and C a finitely generated left Λ-module. We can restate Theorem 10 as follows: *C is projective if, and only if, for every exact sequence* $0 \to A \to B \to C \to 0$ *the derived sequence*

$$0 \to \mathrm{Hom}_\Lambda (C, \Lambda/J) \to \mathrm{Hom}_\Lambda (B, \Lambda/J) \to \mathrm{Hom}_\Lambda (A, \Lambda/J) \to 0$$

is exact. The next exercise provides a companion to this result.

Exercise 5. *Let Λ be a left local ring with radical J and C a finitely generated left Λ-module. Show that the following two statements are equivalent:*

(1) *C is Λ-projective;*

(2) *for every exact sequence $0 \to A \to B \to C \to 0$ of Λ-modules the induced sequence $0 \to A/JA \to B/JB \to C/JC \to 0$ is exact.*

We come now to the result on the global dimension of local rings which was mentioned earlier.

Theorem 11. *Let Λ be a left local ring with radical J. Then, with the usual notation,* $l.\mathrm{GD}\,(\Lambda) = l.\mathrm{Id}_\Lambda (\Lambda/J)$.

Proof. By the definition of the left global dimension† of Λ,

$$l.\mathrm{Id}_\Lambda (\Lambda/J) \leqslant l.\mathrm{GD}\,(\Lambda)$$

---

† See section (3.7).

so it is only necessary to establish the opposite inequality and for this we may suppose that $l.\mathrm{Id}_\Lambda(\Lambda/J) = n$, where $0 \leqslant n < \infty$.

Let C be a finitely generated left Λ-module. By (Chapter 3, Theorem 18), it will suffice to prove that $l.\mathrm{Pd}_\Lambda(C) \leqslant n$.

First suppose that $n = 0$. Then Λ/J is injective. Consequently $\mathrm{Ext}^1_\Lambda(C, \Lambda/J) = 0$ and therefore C is projective by Theorem 10. Thus $l.\mathrm{Pd}_\Lambda(C) \leqslant n$ in this case. Now assume that $n \geqslant 1$. Since

$$l.\mathrm{Id}_\Lambda(\Lambda/J) = n,$$

we can construct an exact sequence

$$0 \to \Lambda/J \to E_0 \to E_1 \to \ldots \to E_n \to 0,$$

where each E_i is injective. On the other hand, because Λ is left Noetherian, we can construct an exact sequence

$$0 \to C_n \to P_{n-1} \to \ldots \to P_1 \to P_0 \to C \to 0,$$

where $P_0, P_1, \ldots, P_{n-1}$ are finitely generated and projective and C_n itself is finitely generated. Evidently the theorem will follow if we show that C_n is projective and this in turn will be established, by virtue of Theorem 10, if we prove that $\mathrm{Ext}^1_\Lambda(C_n, \Lambda/J) = 0$. However, by (Chapter 3, Theorem 14), $\mathrm{Ext}^1_\Lambda(C_n, \Lambda/J)$ is isomorphic to

$$\mathrm{Ext}^1_\Lambda(C, E_n)$$

and this vanishes because E_n is injective. Accordingly the proof of the theorem is complete.

6.4 Local Quasi-Frobenius rings

In this section we return briefly to the study of Quasi-Frobenius rings. These were defined in section (5.5). They make a reappearance because, when one confines one's attention to quasi-local rings, it is possible to characterize them in a new and interesting manner. The next theorem should be compared with Theorem 22 of Chapter 5 from which, in fact, it will be derived.

Theorem 12. *Let Λ be a quasi-local ring such that both Λ_l and Λ_r have finite length. Then the following two statements are equivalent:*

    (a) *the intersection of each pair of non-zero left ideals and each pair of non-zero right ideals is non-zero;*

    (b) *Λ is a Quasi-Frobenius ring.*

178   LOCAL HOMOLOGICAL ALGEBRA

Proof. *Assume* (*a*). The ring Λ is non-trivial and left Artinian, and therefore it contains at least one simple left ideal. However, by condition (*a*), there cannot be two different simple left ideals. Thus there is exactly one simple left ideal I (say) and likewise there is exactly one simple right ideal H (say). *We claim that $I = H$.* For let $\eta \in \Lambda$ and define a mapping $f : I \to I\eta$ by $f(\lambda) = \lambda\eta$. This is a homomorphism of left Λ-modules. Since I is simple, either $f(I)$ is simple or $f(I) = 0$. In either case, $f(I) \subseteq I$ that is $I\eta \subseteq I$. This shows that I is a non-zero right ideal and, since Λ is right Artinian, it contains a simple right ideal. Accordingly $H \subseteq I$ and, by similar arguments, $I \subseteq H$. Thus $I = H$ as claimed.

By (Chapter 5, Theorem 22), we can show that Λ is a Quasi-Frobenius ring by proving that if M is a simple module, then its dual has length at most unity. Without loss of generality we may suppose that M is a left Λ-module in which case it is isomorphic to Λ_l/J. However, by (Chapter 5, Theorem 12), the dual of Λ_l/J is isomorphic to the *right* annihilator J^0 of J. Accordingly we may complete the first part of the proof by showing that J^0 is contained in H.

Let $\lambda \in \Lambda$. Since J is a two-sided ideal, $J(\lambda J^0) = 0$ and therefore $\lambda J^0 \subseteq J^0$. Accordingly J^0 is a two-sided ideal and, because $JJ^0 = 0$, we may regard it as a left module over the division ring Λ/J. Hence, as a left Λ/J-module, J^0 is isomorphic to a direct sum of copies of Λ_l/J and therefore it is a direct sum of simple left ideals. But I is the only simple left ideal. Accordingly $J^0 \subseteq I = H$ and we have proved that (*a*) implies (*b*).

Assume (*b*). Then Λ_l and Λ_r are injective. Now we have a ring isomorphism $\mathrm{End}_\Lambda (\Lambda_r) \approx \Lambda$ in which ϕ in $\mathrm{End}_\Lambda (\Lambda_r)$ corresponds to $\phi(1)$ in Λ. It follows that $\mathrm{End}_\Lambda (\Lambda_r)$ is a quasi-local ring and hence (Chapter 4, Exercise 8) that Λ_r is an indecomposable injective. On the other hand, $\mathrm{End}_\Lambda (\Lambda_l)$ is anti-isomorphic to Λ and therefore it too is quasi-local. Thus Λ_l is an indecomposable injective as well. The fact that condition (*a*) is satisfied now follows from (Chapter 4, Exercise 8).

6.5 Modules over a commutative ring

A good deal is known about the homological properties of *commutative* local rings, but familiarity with the proofs of the main results makes it clear that the full force of the commutative law is not required. In fact one can obtain some interesting generalizations via the notion of what we shall call *semi-commutative* local rings. However before we can win through to these extensions, it will be necessary to give a brief account of certain parts of classical commutative algebra.

In what follows S will always denote a commutative ring with identity element. If A and B are ideals of S, then by AB we mean the set of all finite sums $a_1 b_1 + a_2 b_2 + \ldots + a_m b_m$, where $a_i \in A, b_i \in B$. Of course AB is an ideal of S, and multiplication of ideals is commutative and associative.

Let A be an ideal of S and denote by $\operatorname{Rad} A$ the set composed of all elements of S that have positive powers that are contained in A. This is an ideal. We call $\operatorname{Rad} A$ the *radical* of A, though the reader should be careful not to confuse this notion with that of the radical of a local ring. It is easy to see that if $A \neq S$, then $\operatorname{Rad} A \neq S$. Also if $A_1, A_2, \ldots, A_n$ are ideals of S, then

$$\operatorname{Rad} (A_1 \cap A_2 \cap \ldots \cap A_n) = \operatorname{Rad} A_1 \cap \operatorname{Rad} A_2 \cap \ldots \cap \operatorname{Rad} A_n. \quad (6.5.1)$$

Exercise 6. *Let A and B be ideals of S, where B is finitely generated and $B \subseteq \operatorname{Rad} A$. Show that $B^m \subseteq A$ for some positive integer m.*

We recall that an ideal P of S is called a *prime ideal*, if (i) $P \neq S$, and (ii) when $\alpha, \beta \in S$ and $\alpha\beta \in P$, either $\alpha \in P$ or $\beta \in P$. Thus P is a prime ideal if and only if S/P is an integral domain. Note that if P is a prime ideal and $A_1, A_2, \ldots, A_m$ are ideals such that $A_1 A_2 \ldots A_m \subseteq P$, then $A_i \subseteq P$ for at least one value of i.

The next theorem will find many applications.

Theorem 13. *Let A be an ideal of S and let $P_1, P_2, \ldots, P_m$ be prime ideals. If now A is not contained by any P_i, then there exists $\alpha \in A$ such that $\alpha \notin P_1, \alpha \notin P_2, \ldots, \alpha \notin P_m$.*

Proof. We may assume that none of $P_1, P_2, \ldots, P_m$ is contained by any of the others. Suppose that $1 \leqslant i \leqslant m$. Then

$$AP_1 \ldots P_{i-1} P_{i+1} \ldots P_m$$

is not contained in P_i so there exists α_i in $AP_1 \ldots P_{i-1} P_{i+1} \ldots P_m$ such that $\alpha_i \notin P_i$. Thus $\alpha_i \in A, \alpha_i \notin P_i$ whereas $\alpha_i \in P_j$ if $i \neq j$. Put

$$\alpha = \alpha_1 + \alpha_2 + \ldots + \alpha_m.$$

Then α has the required properties.

Let M be an S-module. We denote by $\operatorname{Ann}_S M$ the ideal formed by all elements $s \in S$ such that $sM = 0$. This ideal is called the *annihilator* of M. Now assume that N is a submodule of M.

Definition. *We say that N is a 'primary submodule' of M if (a) $N \neq M$, and (b) whenever $s \in S, x \in M$ and $sx \in N$, either $x \in N$ or*

$$s \in \mathrm{Rad}\,(\mathrm{Ann}_S\,(M/N)).$$

Thus if N is a primary submodule of M and $sx \in N$, then either $x \in N$ or $s^h M \subseteq N$ for some positive integer h.

Assume that N is a primary submodule of M and put

$$P = \mathrm{Rad}\,(\mathrm{Ann}_S\,(M/N)).$$

Since $M/N \neq 0$, it follows that $P \neq S$. We claim that P is a prime ideal. For suppose that $\alpha, \beta \in S, \alpha\beta \in P$, but $\beta \notin P$. Then there exists an integer h such that $\alpha^h \beta^h M \subseteq N$. But $\beta \notin P$ and therefore $\beta^h \notin \mathrm{Ann}_S\,(M/N)$. Accordingly we can find $x \in M$ such that $\beta^h x \notin N$. However $\alpha^h(\beta^h x) \in N$ and N is a primary submodule of M. It follows that

$$\alpha^h \in \mathrm{Rad}\,(\mathrm{Ann}_S\,(M/N)).$$

Accordingly some power of α^h (and hence some power of α) belongs to $\mathrm{Ann}_S(M/N)$. Thus $\alpha \in P$ and our claim is established. The connection between P, N and M is described by saying that N is a *P-primary* submodule of M.

Exercise 7. *Let P be a prime ideal of S and let $N_1, N_2, ..., N_q\,(q \geqslant 1)$ be P-primary submodules of an S-module M. Show that $N_1 \cap N_2 \cap ... \cap N_q$ is also a P-primary submodule of M.*

Let K be a submodule of an S-module M and Σ a non-empty subset of S. We denote by $K :_M \Sigma$ the submodule of M that consists of all x, in M, such that $\sigma x \in K$ for every σ in Σ. This construction has many useful properties.

First
$$K \subseteq K :_M \Sigma \subseteq M \tag{6.5.2}$$

and if $K_1, K_2, ..., K_t$ are submodules of M, then

$$(K_1 \cap K_2 \cap ... \cap K_t) :_M \Sigma = (K_1 :_M \Sigma) \cap (K_2 :_M \Sigma) \cap ... \cap (K_t :_M \Sigma). \tag{6.5.3}$$

Also if $S\Sigma$ denotes the ideal generated by Σ, then

$$K :_M \Sigma = K :_M S\Sigma \tag{6.5.4}$$

so that, in particular, $K :_M \sigma = K :_M S\sigma$ $(6.5.5)$

for any element σ in S. Another elementary relation is

$$K :_M \Sigma = \bigcap_{\sigma \in \Sigma} (K :_M \sigma) \tag{6.5.6}$$

which is an immediate consequence of the definition. From (6.5.4) and (6.5.6) we obtain

$$K:_M (S\sigma_1 + S\sigma_2 + \ldots + S\sigma_p) = (K:_M \sigma_1) \cap (K:_M \sigma_2) \cap \ldots \cap (K:_M \sigma_p).$$

Finally if A and B are ideals of S, then

$$\text{(6.5.7)}$$

$$(K:_M A):_M B = K:_M AB. \qquad \text{(6.5.8)}$$

Lemma 3. *Let N be a P-primary submodule of the S-module M. If now $\sigma \in S$ but $\sigma \notin P$, then $N:_M \sigma = N$.*

Proof. Suppose that $x \in M$ and $\sigma x \in N$. Since σ is not in

$$\text{Rad}\,(\text{Ann}_S\,(M/N)),$$

we have $x \in N$. Thus $N:_M \sigma \subseteq N$ and the opposite inclusion is obvious.

We must now direct our attention to the theory of primary decompositions. To this end let K be a submodule of an S-module M. We say that K has a *primary decomposition* in M if it can be expressed in the form

$$K = N_1 \cap N_2 \cap \ldots \cap N_q, \qquad \text{(6.5.9)}$$

where each N_i is a primary submodule of M. Suppose that this is the case and that N_i is a P_i-primary submodule of M. The P_i may not all be different. However in this situation we can use Exercise 7 to group together those primary submodules which are associated with a common prime ideal. Thus we can secure that no two of $P_1, P_2, \ldots, P_q$ are the same. Assume that this has been done. If N_i contains

$$N_1 \cap \ldots \cap N_{i-1} \cap N_{i+1} \cap \ldots \cap N_q,$$

then it is superfluous, that is it can be dropped from (6.5.9) without destroying the equation. Supposing for the moment that there exist superfluous N_is, let us take the first one and omit it and then renumber the remaining primary submodules so as to close the gap. If the result still contains superfluous N_is we repeat the operation. And so on. Eventually we arrive at a primary decomposition

$$K = N_1 \cap N_2 \cap \ldots \cap N_h,$$

where (a) the prime ideals $P_i\,(i = 1, 2, \ldots, h)$ are distinct and (b) none of the $N_i\,(i = 1, 2, \ldots, h)$ is superfluous. We then say that

$$K = N_1 \cap N_2 \cap \ldots \cap N_h$$

is a *normal primary decomposition* of K in M. Note that any primary decomposition of K in M can always be refined (by the above procedures) to one that is normal.

Theorem 14. *Let K be a submodule of the S-module M and suppose that K has a primary decomposition in M. Further let*

$$K = N_1 \cap N_2 \cap \ldots \cap N_q$$

and $K = N_1' \cap N_2' \cap \ldots \cap N_t'$ be normal primary decompositions of K in M. If now N_i resp. N_j' is P_i-primary resp. P_j'-primary, then the two sets $\{P_1, P_2, \ldots, P_q\}$ and $\{P_1', P_2', \ldots, P_t'\}$ of prime ideals contain the same members.

Proof. Let P be maximal (with respect to inclusion) among

$$P_1, P_2, \ldots, P_q, \quad P_1', P_2', \ldots, P_t'.$$

Without loss of generality we can suppose that $P = P_t'$. *We claim that P occurs among $P_1, P_2, \ldots, P_q$.* For assume the contrary. By Theorem 13, we can find $s \in P$ so that s is not in any of $P_1, \ldots, P_q, P_1', \ldots, P_{t-1}'$. Since $P = \operatorname{Rad}(\operatorname{Ann}_S(M/N_t'))$ there exists a positive integer h such that $s^h M \subseteq N_t'$ and therefore $N_t' :_M s^h = M$. Suppose that $1 \leqslant i \leqslant q$. Then $s^h \notin P_i$ and therefore, by Lemma 3, $N_i :_M s^h = N_i$. Similarly

$$N_j' :_M s^h = N_j'$$

for $1 \leqslant j \leqslant t-1$. It follows that

$$K :_M s^h = (N_1 :_M s^h) \cap (N_2 :_M s^h) \cap \ldots \cap (N_q :_M s^h)$$
$$= N_1 \cap N_2 \cap \ldots \cap N_q$$
$$= K.$$

On the other hand

$$K :_M s^h = (N_1' :_M s^h) \cap (N_2' :_M s^h) \cap \ldots \cap (N_t' :_M s^h)$$
$$= N_1' \cap \ldots \cap N_{t-1}' \cap M$$
$$= N_1' \cap N_2' \cap \ldots \cap N_{t-1}'.$$

Accordingly $N_1' \cap N_2' \cap \ldots \cap N_{t-1}' = K \subseteq N_t'$ and therefore N_t' is superfluous in the representation $K = N_1' \cap N_2' \cap \ldots \cap N_t'$. This gives a contradiction and thereby establishes our claim. We may therefore suppose that $P_q = P = P_t'$.

Choose (using Theorem 13) $\sigma \in P$ so that σ is not in any of

$$P_1, \ldots, P_{q-1}, P_1', \ldots, P_{t-1}'$$

and let μ be a positive integer so large that $\sigma^\mu M \subseteq N_q$ and $\sigma^\mu M \subseteq N_t'$. Then $N_q :_M \sigma^\mu = M$, $N_t' :_M \sigma^\mu = M$, $N_i :_M \sigma^\mu = N_i\,(1 \leqslant i \leqslant q-1)$, and

$$N_j' :_M \sigma^\mu = N_j'\,(1 \leqslant j \leqslant t-1).$$

Put $K^* = K :_M \sigma^\mu$. Then

$$K^* = N_1 \cap N_2 \cap \ldots \cap N_{q-1} = N_1' \cap N_2' \cap \ldots \cap N_{t-1}'.$$

Here we have two *normal* primary decompositions of K^* in M. The theorem now follows by induction on $p = \max(q, t)$ as soon as we observe that we have to do with a triviality if $p = 1$.

Suppose that M is an S-module and that the submodule K has a primary decomposition in M. Then it has a normal primary decomposition, say

$$K = N_1 \cap N_2 \cap \ldots \cap N_q,$$

where N_i is P_i-primary. By Theorem 14, the prime ideals $P_1, P_2, \ldots, P_q$ are determined solely by K and M, i.e. they are independent of the normal primary decomposition selected. We refer to $P_1, P_2, \ldots, P_q$ as *the prime ideals belonging to the submodule K of M*. In this connection we regard M itself as an *empty* intersection of primary submodules of M. Thus M, considered as a submodule of itself, has a primary decomposition, but the set of prime ideals belonging to M is *empty*.

The discussion of primary decompositions will be continued in the next section. We shall conclude this one by giving preliminary consideration to another topic which also has its origins in commutative algebra.

Let M be an S-module. A sequence $\alpha_1, \alpha_2, \ldots, \alpha_p$ of elements of S is called an *S-sequence on M* if, for each i $(1 \leqslant i \leqslant p)$, α_i is not a zero-divisor on $M/(\alpha_1 M + \alpha_2 M + \ldots + \alpha_{i-1} M)$, that is if

$$(\alpha_1 M + \alpha_2 M + \ldots + \alpha_{i-1} M) :_M \alpha_i = \alpha_1 M + \alpha_2 M + \ldots + \alpha_{i-1} M$$

$$(1 \leqslant i \leqslant p).$$

Thus α by itself is an S-sequence on M if and only if α is not a zero-divisor on M. Again an S-sequence on S (considered as a module with respect to itself) is described simply as an *S-sequence*. This conforms with the terminology introduced in section (3.8).

Let $\alpha_1, \alpha_2, \ldots, \alpha_p$ belong to S, let M be an S-module, and suppose that k satisfies $1 \leqslant k \leqslant p$. Then $\alpha_1, \alpha_2, \ldots, \alpha_p$ is an S-sequence on M if, and only if, $\alpha_1, \alpha_2, \ldots, \alpha_k$ is an S-sequence on M and $\alpha_{k+1}, \alpha_{k+2}, \ldots, \alpha_p$ an S-sequence on $M/(\alpha_1 M + \alpha_2 M + \ldots + \alpha_k M)$. Usually the order of the terms of an S-sequence on M cannot be altered without destroying the S-sequence property.[†] However we do have

Lemma 4. *Suppose that $1 \leqslant i < p$ and let $\alpha_1, \ldots, \alpha_i, \alpha_{i+1}, \ldots, \alpha_p$ be an S-sequence on M. Assume that*

$$(\alpha_1 M + \ldots + \alpha_{i-1} M) :_M \alpha_{i+1} = \alpha_1 M + \alpha_2 M + \ldots + \alpha_{i-1} M.$$

Then $\alpha_1, \ldots, \alpha_{i-1}, \alpha_{i+1}, \alpha_i, \alpha_{i+2}, \ldots, \alpha_p$ is also an S-sequence on M.

† See (21) for more on this point.

Proof. Put $K = \alpha_1 M + \ldots + \alpha_{i-1} M$, $\alpha_i = \alpha$ and $\alpha_{i+1} = \beta$. Then

$$K:_M \alpha = K, \quad K:_M \beta = K \quad \text{and} \quad (K + \alpha M):_M \beta = K + \alpha M.$$

It will suffice to show that $(K + \beta M):_M \alpha = K + \beta M$. To this end assume that $x \in M$ and that αx belongs to $K + \beta M$, say $\alpha x = u + \beta y$, where $u \in K$ and $y \in M$. Then y is in $(K + \alpha M):_M \beta = K + \alpha M$, say $y = v + \alpha z$, where $v \in K$ and $z \in M$. We now have $\alpha(x - \beta z) \in K$. Consequently $x - \beta z$ is in K and therefore $x \in K + \beta M$. Accordingly

$$(K + \beta M):_M \alpha \subseteq K + \beta M$$

and, as the opposite inclusion is obvious, the lemma follows.

Our next result, apart from its intrinsic interest, will be useful later in developing the theory of S-sequences on a module.

Theorem 15. *Let M be an S-module, A an ideal of S, and α an element of A which is not a zerodivisor on M. Then $\{\alpha M:_M A\}/\alpha M$ and*

$$\mathrm{Ext}_S^1 (S/A, M)$$

are isomorphic S-modules.

Proof. The exact sequence $0 \to M \xrightarrow{\alpha} M \to M/\alpha M \to 0$ gives rise to a further exact sequence namely

$$\mathrm{Hom}_S (S/A, M) \xrightarrow{\phi} \mathrm{Hom}_S (S/A, M/\alpha M) \to$$
$$\mathrm{Ext}_S^1 (S/A, M) \xrightarrow{\psi} \mathrm{Ext}_S^1 (S/A, M).$$

Here, since S is its own centre, all the terms are S-modules. Moreover, by (Chapter 3, Exercise 7), all the mappings are S-homomorphisms.

Let U be an S-module then, as we saw in the discussion of (Chapter 2, Exercise 20), $\mathrm{Hom}_S (S/A, U)$ and $0:_U A$ are isomorphic S-modules. Now $0:_M A = 0$ because $\alpha \in A$ and α is not a zerodivisor on M. Accordingly $\mathrm{Hom}_S (S/A, M) = 0$. Again the mapping ψ consists in multiplication by α and $\alpha(S/A) = 0$. Thus ψ is a null homomorphism. This shows that our exact sequence provides an isomorphism

$$\mathrm{Hom}_S (S/A, M/\alpha M) \approx \mathrm{Ext}_S^1 (S/A, M).$$

Finally $\mathrm{Hom}_S (S/A, M/\alpha M)$ is isomorphic, as an S-module, to

$$(0:_{M/\alpha M} A) = (\alpha M:_M A)/\alpha M$$

and now the proof is complete.

6.6 Algebras

We have now to find a fruitful way of linking the results taken from commutative algebra to the theory of non-commutative rings. A convenient way of doing this is to invoke the notion of an *algebra*. To this end, S will always denote a *commutative* ring (with an identity element) and Λ a ring (with an identity element) which need not be commutative.

Let K be a submodule of a Λ-module M. We say that K is an *irreducible submodule* if (a) $K \neq M$, and (b) K is not the intersection of two strictly larger submodules of M.

Exercise 8. *Let M be a Noetherian Λ-module. Show that every submodule of M can be expressed as a finite intersection of irreducible submodules.*

In this exercise it is to be understood that M itself is to be regarded as an *empty* intersection of irreducible submodules of M.

Suppose now that we have a ring-homomorphism $\phi : S \to \Lambda$ (taking identity element into identity element) and that $\phi(S)$ is contained in the centre of Λ. In these circumstances we say that Λ is an *S-algebra* and refer to ϕ as the *structural homomorphism*.

Assume that Λ is an S-algebra and that M is a left Λ-module. Then the structural homomorphism ϕ enables us to regard M and all its Λ-submodules as modules over the ring S. (If $s \in S$ and $x \in M$, we put $sx = \phi(s)x$.) Suppose that K is a Λ-submodule of M and that $s \in S$. Then both sM and $K :_M s$ are Λ-submodules of M and not merely S-submodules.

Let us say that a two-sided ideal I of Λ is a *central ideal* if it can be generated by elements which belong to the centre of Λ. For example, when Λ is an S-algebra and C is an ideal of S, $C\Lambda$ is a central ideal of Λ. Note that, for a left Λ-module M,

$$CM = (C\Lambda)M \qquad (6.6.1)$$

and this is a Λ-submodule of M. If K is a Λ-submodule of M and I a two-sided ideal of Λ, we define $K :_M I$ to be the Λ-submodule of M consisting of all $x \in M$ such that $Ix \subseteq K$. With this notation

$$K :_M C = K :_M C\Lambda \qquad (6.6.2)$$

for every ideal C of S. In particular $K :_M C$ is a Λ-submodule of M.

Theorem 16. *Let Λ be an S-algebra and M a Noetherian left Λ-module. Further let K be a Λ-submodule of M. If now K and M are regarded as S-modules, then K has a primary decomposition in M.*

Proof. In view of Exercise 8, it will suffice to show that if K is an irreducible Λ-submodule of M, then the S-module K is a primary submodule of the S-module M.

Assume the contrary. Then there exist $x \in M$ and $s \in S$ such that $sx \in K$, $x \notin K$ and $s \notin \mathrm{Rad}\,(\mathrm{Ann}_S\,(M/K))$. Thus, for every

$$\mu > 0, \quad s^\mu M \nsubseteq K.$$

Accordingly $K \subset K:_M s$ and $K \subset K + s^\mu M$ the inclusions being strict. Now

$$(K:_M s) \subseteq (K:_M s^2) \subseteq (K:_M s^3) \subseteq \dots$$

is an increasing sequence of Λ-*submodules* of M which, since M is a Noetherian Λ-module, terminates. Hence we can choose μ so that $K:_M s^\mu = K:_M s^{2\mu}$. We shall now obtain a contradiction of the fact that K is an irreducible Λ-submodule of M by showing that

$$K = (K:_M s^\mu) \cap (K + s^\mu M).$$

Assume that x belongs to $(K:_M s^\mu) \cap (K + s^\mu M)$. It will suffice to prove that x belongs to K. Now $x = u + s^\mu y$, where $u \in K$ and $y \in M$. Since $s^\mu x \in K$, it follows that $s^{2\mu} y \in K$ and therefore y belongs to

$$K:_M s^{2\mu} = K:_M s^\mu.$$

Thus $s^\mu y \in K$ and hence $x \in K$ as required.

Theorem 17. *Let Λ be an S-algebra and a left Noetherian ring, let M be a finitely generated left Λ-module and K a Λ-submodule of M. If now A is an ideal of S, then the following statements are equivalent:*

(1) *A is not contained in any prime ideal belonging to K when K is regarded as an S-submodule of M;*

(2) *there exists $\alpha \in A$ such that $K:_M \alpha = K$;*

(3) *$K:_M A = K$.*

Proof. M is a Noetherian Λ-module. Consequently, by Theorem 16, K, when regarded as an S-submodule of M, has a primary decomposition in M. Let $K = N_1 \cap N_2 \cap \dots \cap N_q$, where N_i is a P_i-primary submodule of the S-module M, be a normal primary decomposition. For the rest of the proof all modules are thought of as S-modules.

Assume (1). For each $i\,(1 \leqslant i \leqslant q)$ we have $A \nsubseteq P_i$. Hence, by Theorem 13, there exists $\alpha \in A$ such that $\alpha \notin P_1, \alpha \notin P_2, \dots, \alpha \notin P_q$. By Lemma 3, $N_i:_M \alpha = N_i$. Consequently

$$K:_M \alpha = (N_1:_M \alpha) \cap (N_2:_M \alpha) \cap \dots \cap (N_q:_M \alpha)$$
$$= N_1 \cap N_2 \cap \dots \cap N_q$$
$$= K.$$

Thus (1) implies (2) and it is obvious that (2) implies (3).

Assume (3). Let A_0 be a *finitely generated* S-ideal contained in A. Since Λ is left Noetherian, we can choose A_0 so as to maximize $A_0\Lambda$ in which case it is clear that $A_0\Lambda = A\Lambda$. Further, by (6.6.2),

$$K = K:_M A = K:_M A\Lambda = K:_M A_0\Lambda = K:_M A_0$$

and it is enough to prove that A_0 is not contained by any of

$$P_1, P_2, ..., P_q.$$

Thus from now on we may add the assumption that A itself is finitely generated.

Since $K = K:_M A$, we see that

$$K = (K:_M A):_M A = K:_M A^2$$

and, repeating the argument, it follows that $K:_M A^\mu = K$ for all $\mu > 0$.

Renumber the N_i so that $A \nsubseteq P_1, A \nsubseteq P_2, ..., A \nsubseteq P_h$ but

$$A \subseteq P_{h+1}, A \subseteq P_{h+2}, ..., A \subseteq P_q.$$

We wish to show that $h = q$. If $h + 1 \leqslant j \leqslant q$, then

$$A \subseteq \operatorname{Rad}(\operatorname{Ann}_S(M/N_j))$$

and A is finitely generated. Hence, by Exercise 6, $A^\mu \subseteq \operatorname{Ann}_S(M/N_j)$ if μ is large enough. Choose μ so that $A^\mu M \subseteq N_j (h+1 \leqslant j \leqslant q)$ in which case $N_j:_M A^\mu = M$ for all j between $h+1$ and q. Since P_i is prime, $A^\mu \nsubseteq P_i$ provided that $1 \leqslant i \leqslant h$. For such an i we can select $\alpha_i \in A^\mu$ so that $\alpha_i \notin P_i$ and then, by Lemma 3,

$$N_i \subseteq N_i:_M A^\mu \subseteq N_i:_M \alpha_i = N_i$$

showing that $N_i:_M A^\mu = N_i$. Accordingly

$$K:_M A^\mu = (N_1:_M A^\mu) \cap ... \cap (N_h:_M A^\mu) \cap (N_{h+1}:_M A^\mu) \cap ... \cap (N_q:_M A^\mu)$$
$$= N_1 \cap N_2 \cap ... \cap N_h.$$

But $K = K:_M A^\mu$. Thus $K = N_1 \cap N_2 \cap ... \cap N_h$ and now we see that $h = q$ for otherwise N_q would have been superfluous in the original decomposition.

We return now to the study of S-sequences on a module.

Theorem 18. *Let Λ be an S-algebra and a left Noetherian ring, let M be a finitely generated left Λ-module, and A an ideal of S. Further let $\alpha_1, \alpha_2, ..., \alpha_t$ and $\alpha_1', \alpha_2', ..., \alpha_t'$ be S-sequences on M, where each α_i and α_j' is in A. Then*

$$(\alpha_1 M + \alpha_2 M + ... + \alpha_t M):_M A = \alpha_1 M + \alpha_2 M + ... + \alpha_t M$$

if and only if

$$(\alpha_1' M + \alpha_2' M + \ldots + \alpha_t' M) :_M A = \alpha_1' M + \alpha_2' M + \ldots + \alpha_t' M.$$

Proof. Put

$$U_i = \alpha_1 M + \alpha_2 M + \ldots + \alpha_i M \quad \text{and} \quad U_i' = \alpha_1' M + \alpha_2' M + \ldots + \alpha_i' M$$

for $0 \leqslant i \leqslant t - 1$, where by U_0 and U_0' we mean the zero submodule of M. Then $U_i :_M A = U_i$ because $U_i :_M \alpha_{i+1} = U_i$ and likewise

$$U_i' :_M A = U_i'.$$

Thus if we take the whole collection of prime ideals that belong to the submodules $U_0, \ldots, U_{t-1}, U_0', \ldots, U_{t-1}'$ of M, then, by Theorem 17, A is not contained in any of them. It follows, by Theorem 13, that there exists $\beta \in A$ such that β is not in any of these prime ideals. Consequently $U_i :_M \beta = U_i$ and $U_i' :_M \beta = U_i'$ for $0 \leqslant i \leqslant t - 1$. We see now that $\alpha_1, \ldots, \alpha_{t-1}, \beta$ is an S-sequence on M and, by repeated applications of Lemma 4, conclude that $\beta, \alpha_1, \ldots, \alpha_{t-1}$ is an S-sequence on M. Similar considerations show that both $\alpha_1', \ldots, \alpha_{t-1}', \beta$ and

$$\beta, \alpha_1', \ldots, \alpha_{t-1}'$$

are S-sequences on M.

Neither α_t nor β is a zerodivisor on M/U_{t-1}. Consequently, using Theorem 15, we have S-isomorphisms

$$\{(\alpha_1 M + \ldots + \alpha_{t-1} M + \alpha_t M) :_M A\}/(\alpha_1 M + \ldots + \alpha_{t-1} M + \alpha_t M)$$

$$\approx \{\alpha_t (M/U_{t-1}) :_{M/U_{t-1}} A\}/\alpha_t (M/U_{t-1})$$

$$\approx \operatorname{Ext}_S^1(S/A, M/U_{t-1})$$

$$\approx \{\beta(M/U_{t-1}) :_{M/U_{t-1}} A\}/\beta(M/U_{t-1})$$

$$\approx \{(\alpha_1 M + \ldots + \alpha_{t-1} M + \beta M) :_M A\}/(\alpha_1 M + \ldots + \alpha_{t-1} M + \beta M).$$

It follows that

$$(\alpha_1 M + \ldots + \alpha_{t-1} M + \alpha_t M) :_M A = \alpha_1 M + \ldots + \alpha_{t-1} M + \alpha_t M$$

if and only if

$$(\beta M + \alpha_1 M + \ldots + \alpha_{t-1} M) :_M A = \beta M + \alpha_1 M + \ldots + \alpha_{t-1} M.$$

Likewise

$$(\alpha_1' M + \ldots + \alpha_{t-1}' M + \alpha_t' M) :_M A = \alpha_1' M + \ldots + \alpha_{t-1}' M + \alpha_t' M$$

if and only if

$$(\beta M + \alpha_1' M + \ldots + \alpha_{t-1}' M) :_M A = \beta M + \alpha_1' M + \ldots + \alpha_{t-1}' M.$$

Accordingly it suffices to prove the theorem for the sequences

$$\beta, \alpha_1, ..., \alpha_{t-1} \quad \text{and} \quad \beta, \alpha_1', ..., \alpha_{t-1}'.$$

Thus unless the sequences $\alpha_1, \alpha_2, ..., \alpha_t$ and $\alpha_1', \alpha_2', ..., \alpha_t'$ are identical, we can always increase the number of initial terms they have in common. By repeated applications of this observation we can reduce the theorem to the case where $\alpha_i = \alpha_i'$ for $i = 1, 2, ..., t$ in which situation it is obvious.

Corollary. *Let Λ, S, A and M satisfy the hypotheses of Theorem 18 and suppose that $\alpha_1, \alpha_2, ..., \alpha_t$ and $\alpha_1', \alpha_2', ..., \alpha_\nu'$ are S-sequences on M composed of elements of A. If now $\nu < t$, then it is possible to find $\alpha_{\nu+1}', \alpha_{\nu+2}', ..., \alpha_t'$ in A so that $\alpha_1', ..., \alpha_\nu', \alpha_{\nu+1}', ..., \alpha_t'$ is an S-sequence on M.*

Proof. We have $(\alpha_1 M + ... + \alpha_\nu M):_M A = \alpha_1 M + ... + \alpha_\nu M$ because $(\alpha_1 M + ... + \alpha_\nu M):_M \alpha_{\nu+1} = \alpha_1 M + ... + \alpha_\nu M$. Hence, by Theorem 18, $(\alpha_1' M + ... + \alpha_\nu' M):_M A = \alpha_1' M + ... + \alpha_\nu' M$. It follows, by Theorem 17, that there exists $\alpha_{\nu+1}' \in A$ so that

$$(\alpha_1' M + ... + \alpha_\nu' M):_M \alpha_{\nu+1}' = \alpha_1' M + ... + \alpha_\nu' M.$$

This secures that $\alpha_1', ..., \alpha_\nu', \alpha_{\nu+1}'$ is an S-sequence on M. If now

$$\nu + 1 < t,$$

then we can repeat this argument, and so on until the desired result is obtained.

Let Λ be an S-algebra and a left Noetherian ring, M a finitely generated left Λ-module, and A an S-ideal such that $AM \neq M$. Let us attempt to build up, step by step, an S-sequence $\alpha_1, \alpha_2, \alpha_3 ...$ on M composed of elements of A. *We claim that we must reach a point where we can go no further.* For otherwise we would generate an infinite sequence and the $\alpha_1 M + \alpha_2 M + ... + \alpha_i M$ ($i = 0, 1, 2, ...$) would constitute an increasing sequence of Λ-submodules of M. But M is a Noetherian Λ-module. Accordingly there would exist an integer i such that

$$\alpha_1 M + \alpha_2 M + ... + \alpha_i M = \alpha_1 M + \alpha_2 M + ... + \alpha_i M + \alpha_{i+1} M.$$

Thus $\alpha_1 M + \alpha_2 M + ... + \alpha_i M = (\alpha_1 M + \alpha_2 M + ... + \alpha_i M):_M \alpha_{i+1}$
$$= (\alpha_1 M + \alpha_2 M + ... + \alpha_{i+1} M):_M \alpha_{i+1}$$
$$= M$$

which contradicts our assumption that $AM \neq M$. Thus our claim is established. Let α_t be the last term in the sequence. We then say that

$\alpha_1, \alpha_2, \ldots, \alpha_t$ is a *maximal S-sequence on M in A*. Note the fact that we cannot go any further means that

$$(\alpha_1 M + \alpha_2 M + \ldots + \alpha_t M) :_M A \neq \alpha_1 M + \alpha_2 M + \ldots + \alpha_t M.$$

Again, by Theorem 18 Cor., any two maximal S-sequences on M in A contain the same number of terms. This number will be called the *grade of A on M* and denoted by $\mathrm{gr}\{A; M\}$.

Let Λ, S, M and A be as before so that, in particular, $AM \neq M$ and $\mathrm{gr}\{A; M\}$ is defined. Now let A' be any S-ideal such that

$$A'\Lambda = A\Lambda.$$

Since $\qquad AM = (A\Lambda) M = (A'\Lambda) M = A'M,$

it follows that $A'M \neq M$ and $\mathrm{gr}\{A'; M\}$ is defined. *We claim that*

$$\mathrm{gr}\{A; M\} = \mathrm{gr}\{A'; M\}. \tag{6.6.3}$$

To see this we first note that, by replacing A' by $A + A'$, we may suppose that $A \subseteq A'$. Let $\alpha_1, \alpha_2, \ldots, \alpha_t$ be a maximal S-sequence on M in A. Since

$$\begin{aligned}
(\alpha_1 M + \ldots + \alpha_t M) :_M A &= (\alpha_1 M + \ldots + \alpha_t M) :_M A\Lambda \\
&= (\alpha_1 M + \ldots + \alpha_t M) :_M A'\Lambda \\
&= (\alpha_1 M + \ldots + \alpha_t M) :_M A',
\end{aligned}$$

we see that $(\alpha_1 M + \ldots + \alpha_t M) :_M A' \neq \alpha_1 M + \ldots + \alpha_t M$ and therefore $\alpha_1, \alpha_2, \ldots, \alpha_t$ is also a maximal S-sequence on M in A'. This establishes (6.6.3).

If I_1, I_2 are two-sided ideals of Λ, then we can define their product $I_1 I_2$ just as we did in the case of ideals in a commutative ring. $I_1 I_2$ is also a two-sided ideal and our multiplication is associative but not commutative. With its aid we can give a useful extension of (6.6.3). This is provided by

Exercise 9. *Let Λ be a left Noetherian ring and an S-algebra. Further let M be a finitely generated left Λ-module and A, B ideals of S such that $AM \neq M$ and $BM \neq M$. Show that if $A\Lambda$ contains a power of $B\Lambda$, then $\mathrm{gr}\{B; M\} \leqslant \mathrm{gr}\{A; M\}$.*

Finally assume once more that Λ is a left Noetherian ring and an S-algebra, and that M is a finitely generated left Λ-module. If now I is a two-sided ideal of Λ such that $IM \neq M$, we shall put

$$\mathrm{gr}_S\{I; M\} = \mathrm{gr}\{\phi^{-1}(I); M\}, \tag{6.6.4}$$

where $\phi:S\to\Lambda$ is the structural homomorphism. For example, if there is an S-ideal A such that $I = A\Lambda$, then $A\Lambda = \phi^{-1}(I)\,\Lambda$ and therefore $\mathrm{gr}_S\{I; M\} = \mathrm{gr}\{A; M\}$ by virtue of (6.6.3).

6.7 Semi-commutative local algebras

As before S will denote a commutative ring with an identity element. Let Λ be a quasi-local S-algebra with radical J.

Definition. *The quasi-local S-algebra Λ will be said to be 'semi-commutative' if there exists an S-ideal A such that $J^k \subseteq A\Lambda \subseteq J$ for some positive integer k. It will be said to be 'strongly semi-commutative' if the ideal A can be chosen so that $A\Lambda = J$.*

Observe that if Λ is a quasi-local S-algebra and $I \neq \Lambda$ is a two-sided ideal of Λ, then I is contained in the radical J. Also Λ/I has an obvious structure as a quasi-local S-algebra and its radical is J/I. Furthermore if Λ is semi-commutative respectively strongly semi-commutative, then Λ/I is semi-commutative respectively strongly semi-commutative.

Now suppose that Λ is a left local ring and a semi-commutative S-algebra. Further let $M \neq 0$ be a finitely generated left Λ-module. By Theorem 4, $JM \neq M$ and therefore $\mathrm{gr}_S\{J; M\}$ is defined. Let A be *any* S-ideal such that $J^k \subseteq A\Lambda \subseteq J$ and put $\phi^{-1}(J) = B$, where $\phi:S\to\Lambda$ is the structural homomorphism. Then $\mathrm{gr}_S\{J, M\} = \mathrm{gr}\{B; M\}$ and $J^k \subseteq A\Lambda \subseteq B\Lambda \subseteq J$. Accordingly each of $A\Lambda$ and $B\Lambda$ contains a power of the other and therefore, by Exercise 9, $\mathrm{gr}\{A; M\} = \mathrm{gr}\{B; M\}$. It follows that

$$\mathrm{gr}_S\{J; M\} = \mathrm{gr}\{A; M\}. \tag{6.7.1}$$

This observation will be used frequently in the sequel.

Lemma 5. *Let Λ be a left local ring with radical J and a semi-commutative S-algebra. If $\mathrm{gr}_S\{J; \Lambda_l\} = 0$, then there exists $\lambda \in \Lambda$ such that $\lambda \neq 0$ but $\lambda J = 0$.*

Proof. Choose an ideal A, of S, and a positive integer k such that $J^k \subseteq A\Lambda \subseteq J$. Then, by (6.7.1), $\mathrm{gr}\{A; \Lambda_l\} = 0$ and therefore every element of A is a zerodivisor on Λ_l. It follows, from Theorem 17, that $0:_{\Lambda_l} A \neq 0$ and therefore there exists $\lambda' \in \Lambda_l$ such that $\lambda' \neq 0$ and $A\lambda' = 0$. Thus $\lambda'A = 0$ and therefore $\lambda'(A\Lambda) = 0$. This shows that $\lambda'J^k = 0$. Choose the smallest integer h such that $\lambda'J^h = 0$. Then $h \geqslant 1$ and $\lambda'J^{h-1} \neq 0$. Select $\lambda'' \in J^{h-1}$ so that $\lambda'\lambda'' \neq 0$. Then $\lambda'\lambda''J = 0$ and the proof is complete.

Our next result generalizes an important theorem in the homology theory of commutative Noetherian rings.

Theorem 19. *Let Λ be a left local ring with radical J and a semicommutative S-algebra. Further let $M \neq 0$ be a finitely generated left Λ-module such that $l.\mathrm{Pd}_\Lambda(M) < \infty$. Then*

$$\mathrm{gr}_S\{J; M\} + l.\mathrm{Pd}_\Lambda(M) = \mathrm{gr}_S\{J; \Lambda_l\},$$

where J is the radical of Λ.

Proof. Choose an S-ideal A such that $J^k \subseteq A\Lambda \subseteq J$ for a suitable integer k. Thus $\mathrm{gr}_S\{J; \Lambda_l\} = \mathrm{gr}\{A; \Lambda_l\} = n$ say. We shall use induction on n.

If $n = 0$ then, by Lemma 5, there exists $\lambda \in \Lambda$ such that $\lambda \neq 0$ but $\lambda J = 0$. Hence, by Theorem 9, M is a non-zero free module. Since $\mathrm{gr}\{A; \Lambda_l\} = 0$, every element of A is a zerodivisor on Λ_l and therefore also a zerodivisor on M. Accordingly $\mathrm{gr}_S\{J; M\} = \mathrm{gr}\{A; M\} = 0$ and we see that the theorem holds for the case $n = 0$. We shall now suppose that $n > 0$ and that the theorem has been proved for smaller values of the inductive variable. Since $\mathrm{gr}\{A; \Lambda_l\} = n > 0$, there exists $\alpha \in A$ such that α is not a zerodivisor on Λ.

First assume that $\mathrm{gr}_S\{J; M\} = 0$, i.e. that $\mathrm{gr}\{A; M\} = 0$. Let (F, ψ) be a projective cover for the Λ-module M and construct an exact sequence $0 \to K \to F \overset{\psi}{\to} M \to 0$, where $K = \mathrm{Ker}\,\psi$. Then K is a finitely generated Λ-module and, by Theorem 5, $K \subseteq JF$. Further, by the same theorem, F is a free module and therefore α is not a zerodivisor on F or K. Let $\phi: S \to \Lambda$ be the structural homomorphism. Then $\phi(\alpha)$ belongs to J and to the centre of Λ, and it is not a zerodivisor on either Λ or K. We may therefore apply Theorem 8 and so obtain

$$l.\mathrm{Pd}_{\Lambda/\alpha\Lambda}(K/\alpha K) = l.\mathrm{Pd}_\Lambda(K) < \infty$$

since $l.\mathrm{Pd}_\Lambda(M) < \infty$. Now $\mathrm{gr}\{A; M\} = 0$ and therefore α is a zerodivisor on M. This shows that M is not free and hence not projective (Theorem 6). Thus $K \neq 0$ and $l.\mathrm{Pd}_\Lambda(K) = l.\mathrm{Pd}_\Lambda(M) - 1$. Accordingly

$$l.\mathrm{Pd}_\Lambda(M) = l.\mathrm{Pd}_{\Lambda/\alpha\Lambda}(K/\alpha K) + 1.$$

Consider $\Lambda/\alpha\Lambda$. This is a left local ring with radical $J/\alpha\Lambda$, and it is also a semi-commutative S-algebra. Further

$$(J/\alpha\Lambda)^k \subseteq A(\Lambda/\alpha\Lambda) \subseteq J/\alpha\Lambda$$

and $$\mathrm{gr}_S\{J/\alpha\Lambda; \Lambda_l/\alpha\Lambda_l\} = \mathrm{gr}\{A; \Lambda_l/\alpha\Lambda_l\} = n - 1.$$

We may therefore apply the inductive hypothesis to the $\Lambda/\alpha\Lambda$-module $K/\alpha K$. This shows that

$$l.\mathrm{Pd}_{\Lambda/\alpha\Lambda}\,(K/\alpha K)+\mathrm{gr}\,\{A\,;K/\alpha K\}=n-1$$

and therefore $l.\mathrm{Pd}_\Lambda\,(M)+\mathrm{gr}\,\{A\,;K/\alpha K\}=n.$ (6.7.2)

Now $\mathrm{gr}\,\{A\,;M\}=0$ and therefore, by Theorem 17, there exists $x\in M$ such that $x\neq 0$ but $Ax=0$. Choose $\xi\in F$ so that $\psi(\xi)=x$. Then $A\xi\subseteq\mathrm{Ker}\,\psi=K$ and hence, in particular, $\alpha\xi\in K$. However $\alpha\xi\notin\alpha K$ because $\xi\notin K$ and α is not a zerodivisor on F. But $A(\alpha\xi)=\alpha(A\xi)\subseteq\alpha K$ and therefore $\alpha K:_K A\neq\alpha K$. This shows that $\mathrm{gr}\,\{A\,;K/\alpha K\}=0$. Accordingly, by (6.7.2), $l.\mathrm{Pd}_\Lambda\,(M)=n=\mathrm{gr}_S\{J\,;\Lambda_l\}$ and the inductive step has been taken when $\mathrm{gr}_S\{J\,;M\}=0$.

Now suppose that $\mathrm{gr}_S\{J\,;M\}>0$. We can choose an element α, in A, not only so that it is not a zerodivisor on Λ but also so that it is not a zerodivisor on M. Then, by Theorem 8,

$$l.\mathrm{Pd}_{\Lambda/\alpha\Lambda}\,(M/\alpha M)=l.\mathrm{Pd}_\Lambda\,(M)<\infty$$

and therefore, this time by (Chapter 3, Theorem 21),

$$l.\mathrm{Pd}_\Lambda\,(M/\alpha M)=l.\mathrm{Pd}_\Lambda\,(M)+1.$$

Further, since

$$\begin{aligned}\mathrm{gr}_S\{J\,;M/\alpha M\}&=\mathrm{gr}\,\{A\,;M/\alpha M\}\\&=\mathrm{gr}\,\{A\,;M\}-1\\&=\mathrm{gr}_S\{J\,;M\}-1,\end{aligned}$$

it follows that

$$l.\mathrm{Pd}_\Lambda\,(M/\alpha M)+\mathrm{gr}_S\{J\,;M/\alpha M\}=l.\mathrm{Pd}_\Lambda\,(M)+\mathrm{gr}_S\{J\,;M\}.$$

Hence, in our present situation, we can reduce the value of $\mathrm{gr}_S\{J\,;M\}$ step by step until all we have to do is to establish the desired result when $\mathrm{gr}_S\{J\,;M\}=0$. However this special case has already been covered. Thus the inductive step has been accomplished in full and therefore the proof is complete.

Theorem 20. *Let Λ be an S-algebra and also a left local ring with radical J. Suppose that there exists an S-sequence $\alpha_1,\alpha_2,...,\alpha_q\,(q\geqslant 0)$ on Λ such that $J=\alpha_1\Lambda+\alpha_2\Lambda+...+\alpha_q\Lambda$. Then, with the usual notation for global dimension, $l.\mathrm{GD}\,(\Lambda)=q$.*

Remarks. We regard an *empty sequence* of elements of S as forming an S-sequence on every S-module. Thus the case $q=0$ asserts that if $J=0$, then $l.\mathrm{GD}\,(\Lambda)=0$ which, of course, is obvious because Λ is then a division ring. Note that the conditions of the theorem ensure that Λ is a *strongly* semi-commutative local S-algebra.

Proof. Let $\phi: S \to \Lambda$ be the structural homomorphism. Then

$$\phi(\alpha_1), \phi(\alpha_2), \ldots, \phi(\alpha_q)$$

is a Λ-sequence in the sense of section (3.8),

$$\phi(\alpha_1)\Lambda + \phi(\alpha_2)\Lambda + \ldots + \phi(\alpha_q)\Lambda = J,$$

and $J \neq \Lambda$. Accordingly $l.\mathrm{Pd}_\Lambda(\Lambda/J) = q$ and therefore $l.\mathrm{GD}(\Lambda) \geqslant q$.

We must now prove the opposite inequality. Let M be a finitely generated left Λ-module. Then, by (Chapter 3, Theorem 18), it is enough to show that $l.\mathrm{Pd}_\Lambda(M) \leqslant q$. For this step we shall use induction on q. Note that if $q = 0$ then Λ is a division ring and there is, in this case, no problem.

From here on we assume that $q \geqslant 1$ and make the obvious induction hypothesis. By applying this hypothesis to the ring $\Lambda/\alpha_1\Lambda$ we see that $l.\mathrm{GD}(\Lambda/\alpha_1\Lambda) = q-1$. Since we wish to show that $l.\mathrm{Pd}_\Lambda(M) \leqslant q$ we may suppose that M is not projective. Construct an exact sequence $0 \to K \to F \to M \to 0$, where F is a free module on a finite base and K is a submodule of F. Then K is non-zero and finitely generated. Further $\phi(\alpha_1)$ is not a zerodivisor on F and therefore it is not a zerodivisor on K. By Theorem 8,

$$q - 1 \geqslant l.\mathrm{Pd}_{\Lambda/\alpha_1\Lambda}(K/\alpha_1 K) = l.\mathrm{Pd}_\Lambda(K) = l.\mathrm{Pd}_\Lambda(M) - 1.$$

Thus $l.\mathrm{Pd}_\Lambda(M) \leqslant q$ and the proof is complete.

Our final theorem provides, among other information, a converse to Theorem 20. In order not to obscure the main lines of the argument certain technicalities have been concentrated in the following lemma.

Lemma 6. *Let Λ be a left local ring with radical J and a strongly semi-commutative S-algebra. Suppose that $\mathrm{gr}_S\{J; \Lambda_l\} > 0$. Then there exists $\alpha \in S$ such that α is not a zerodivisor on Λ, $\alpha\Lambda \subseteq J$ and $\alpha\Lambda \nsubseteq J^2$.*

Proof. Let $\phi: S \to \Lambda$ be the structural homomorphism. Since

$$\mathrm{gr}_S\{J; \Lambda_l\} > 0,$$

we have $J \neq 0$ and therefore, by Theorem 4, $J^2 \neq J$. Put $A = \phi^{-1}(J)$ and $B = \phi^{-1}(J^2)$. Since Λ is strongly semi-commutative, $A\Lambda = J$ and, since $J^2 \neq J$, we have $A^2 \subseteq B \subset A$, where the second inclusion is strict.

Consider Λ_l as an S-module. Certain prime ideals $P_1, P_2, \ldots, P_t$ of S will belong to its zero submodule. By renumbering, if necessary, we may arrange that $P_1, P_2, ..., P_\nu$ are the maximal members (with respect to inclusion) of the set $\{P_1, P_2, ..., P_t\}$. Choose $\alpha' \in A$ so that $\alpha' \notin B$. Then α' will belong to some of $P_1, P_2, ..., P_\nu$ but not to others. We shall suppose that $\alpha' \notin P_1, ..., \alpha' \notin P_\mu$ and $\alpha' \in P_{\mu+1}, ..., \alpha' \in P_\nu$.

Since $\mathrm{gr}\{A; \Lambda_l\} = \mathrm{gr}_S\{J; \Lambda_l\} > 0$ it follows that A and hence A^2 is not contained in any of $P_1, P_2, ..., P_t$. Accordingly $B \nsubseteq P_1, ..., B \nsubseteq P_\nu$ and therefore $BP_1 P_2 ... P_\mu$ is not contained by any of $P_{\mu+1}, P_{\mu+2}, ..., P_\nu$. Theorem 13 now shows that there exists $\beta \in BP_1 P_2 ... P_\mu$ such that $\beta \notin P_{\mu+1}, ..., \beta \notin P_\nu$. Put $\alpha = \alpha' + \beta$. Then $\alpha \in A, \alpha \notin B$ and α is not in any of $P_1, P_2, ..., P_\nu$. It follows that α is not a zerodivisor on $\Lambda, \alpha\Lambda \subseteq J$, but $\alpha\Lambda \nsubseteq J^2$. This completes the proof.

The next theorem provides an extension to one of the most important theorems in the homology theory of commutative Noetherian rings.

Theorem 21. *Let Λ be a left local ring with radical J, and suppose that Λ is a strongly semi-commutative S-algebra. Then the following statements are equivalent*:

(a) $l.\mathrm{GD}(\Lambda) < \infty$;

(b) $l.\mathrm{Pd}_\Lambda(\Lambda/J) < \infty$;

(c) $l.\mathrm{Id}_\Lambda(\Lambda/J) < \infty$;

(d) *there is an S-sequence $\alpha_1, \alpha_2, ..., \alpha_q$ on Λ such that*

$$\alpha_1\Lambda + \alpha_2\Lambda + ... + \alpha_q\Lambda = J.$$

Moreover, when these conditions are satisfied,

$$l.\mathrm{GD}(\Lambda) = l.\mathrm{Pd}_\Lambda(\Lambda/J) = l.\mathrm{Id}_\Lambda(\Lambda/J)$$

and their common value is the number of terms in any S-sequence on Λ which generates J.

Proof. In view of Theorems 11 and 20, it will suffice to prove the following assertion: *if $l.\mathrm{Pd}_\Lambda(\Lambda/J) = q$, then there exists an S-sequence $\alpha_1, \alpha_2, ..., \alpha_q$ on Λ such that $\alpha_1\Lambda + \alpha_2\Lambda + ... + \alpha_q\Lambda = J$.* We concentrate on this.

Put $\mathrm{gr}\{J; \Lambda_l\} = t$ and let $\phi: S \to \Lambda$ be the structural homomorphism. Then $J = A\Lambda$, where $A = \phi^{-1}(J)$. We shall use induction on t.

First suppose that $t = 0$. By Theorem 19, $l.\mathrm{Pd}_\Lambda(\Lambda/J) = 0$. Thus Λ/J is a projective Λ-module and therefore, by Theorem 6, it is free. But $J(\Lambda/J) = 0$. Consequently $J = 0$ and the key assertion has been proved for the case $t = 0$.

Now assume that $t > 0$ and that the assertion in italics at the beginning of the proof has been established for all smaller values of the inductive variable. By Lemma 6, there exists $\alpha \in S$ such that α is not a zerodivisor on Λ, $\alpha\Lambda \subseteq J$, and $\alpha\Lambda \nsubseteq J^2$.

Put $\Lambda^* = \Lambda/\alpha\Lambda$ and $J^* = J/\alpha\Lambda$. Then Λ^* is a left local ring with radical J^*, and a strongly semi-commutative S-algebra. Also

$$\mathrm{gr}_S\{J^*; \Lambda_l^*\} = \mathrm{gr}\{A; \Lambda_l/\alpha\Lambda_l\}$$
$$= \mathrm{gr}\{A; \Lambda_l\} - 1$$
$$= t - 1.$$

We claim that J^* is isomorphic (as a left Λ^*-module) to a direct summand of $J/\alpha J$. For $\phi(\alpha) \in J$, $\phi(\alpha) \notin J^2$. Consequently, by Exercise 2, there exist $x_2, x_3, \ldots, x_p$ in J such that $\phi(\alpha), x_2, \ldots, x_p$ is a minimal set of generators for J considered as a left Λ-module. Put

$$U = J\phi(\alpha) + \Lambda x_2 + \ldots + \Lambda x_p.$$

Then $J = \Lambda\phi(\alpha) + U$. Now suppose that $y \in \Lambda\phi(\alpha) \cap U$. In these circumstances $y = \lambda\phi(\alpha) = \omega\phi(\alpha) + \lambda_2 x_2 + \ldots + \lambda_p x_p$, where $\lambda, \lambda_2, \ldots, \lambda_p$ are in Λ and $\omega \in J$. We now see that $(\omega - \lambda)\phi(\alpha) + \lambda_2 x_2 + \ldots + \lambda_p x_p = 0$ whence $\omega - \lambda \in J$ and therefore $\lambda \in J$. Accordingly $y \in J\phi(\alpha)$. It follows that $\Lambda\phi(\alpha) \cap U = J\phi(\alpha)$ and hence that

$$J/\alpha J = J/J\phi(\alpha) = (\Lambda\phi(\alpha)/J\phi(\alpha)) \oplus (U/J\phi(\alpha)).$$

But $\quad J^* = J/\alpha\Lambda = J/\Lambda\phi(\alpha) = (J/J\phi(\alpha))/(\Lambda\phi(\alpha)/J\phi(\alpha)).$

It follows that J^* and $U/J\phi(\alpha)$ are isomorphic as left Λ-modules and hence as left Λ^*-modules. This establishes our claim.

Since α is not a zerodivisor on Λ or J, Theorem 8 shows that

$$l.\mathrm{Pd}_{\Lambda^*}(J/\alpha J) = l.\mathrm{Pd}_\Lambda(J).$$

But $l.\mathrm{Pd}_\Lambda(J) < \infty$ because we have an exact sequence

$$0 \to J \to \Lambda \to \Lambda/J \to 0$$

and we are given that $l.\mathrm{Pd}_\Lambda(\Lambda/J) < \infty$. Hence it follows, by (Chapter 3, Exercise 10), that $l.\mathrm{Pd}_{\Lambda^*}(J^*) \leqslant l.\mathrm{Pd}_{\Lambda^*}(J/\alpha J) < \infty$ and so we see that $l.\mathrm{Pd}_{\Lambda^*}(\Lambda^*/J^*) < \infty$. Accordingly we can apply the inductive hypothesis to Λ^* and so conclude that there exists an S-sequence $\alpha_2, \alpha_3, \ldots, \alpha_h$ on $\Lambda^* = \Lambda/\alpha\Lambda$ such that $J^* = \alpha_2\Lambda^* + \alpha_3\Lambda^* + \ldots + \alpha_h\Lambda^*$. But now $\alpha, \alpha_2, \alpha_3, \ldots, \alpha_h$ is an S-sequence on Λ and

$$J = \alpha\Lambda + \alpha_2\Lambda + \ldots + \alpha_h\Lambda.$$

Finally, by (Chapter 3, Theorem 22), $h = l.\mathrm{Pd}_\Lambda\,(\Lambda/J) = q$ and with this the proof is complete.

It remains for us to give examples of non-commutative local rings to which our theory is applicable. Suppose then that D is a division ring and that $x_1, x_2, \ldots, x_q$ are indeterminates. Then, as was explained in the remarks immediately following Exercise 4, the power series ring $D[[x_1, x_2, \ldots, x_q]]$ is a left and right local ring whose radical J is the two-sided ideal generated by the indeterminates $x_1, x_2, \ldots, x_q$ themselves.

Now let $y_1, y_2, \ldots, y_q$ be further indeterminates and use Z to denote the ring of integers. There is a ring-homomorphism ϕ which maps the polynomial ring $Z[y_1, y_2, \ldots, y_q]$ into $D[[x_1, x_2, \ldots, x_q]]$ in such a way that $\phi(y_i) = x_i$ for $1 \leqslant i \leqslant q$. As a result $D[[x_1, x_2, \ldots, x_q]]$ becomes a strongly semi-commutative $Z[y_1, y_2, \ldots, y_q]$-algebra. Finally if I is a two-sided ideal of $D[[x_1, x_2, \ldots, x_q]]$ which is different from the ring itself, then $\Lambda = D[[x_1, x_2, \ldots, x_q]]/I$ is a left and right local ring and a strongly semi-commutative S-algebra, where $S = Z[y_1, y_2, \ldots, y_q]$. Thus Λ is a ring to which the theorems of this section are applicable and, of course, it will (in general) be non-commutative.

Solutions to the Exercises on Chapter 6

Exercise 1. *Let Λ be a quasi-local ring with radical J, M a finitely generated left Λ-module, and $u_1, u_2, \ldots, u_n$ elements of M. Denote by $\bar{u}_i$ the natural image of u_i in M/JM. Show that $u_1, u_2, \ldots, u_n$ generate M if and only if $\bar{u}_1, \bar{u}_2, \ldots, \bar{u}_n$ generate M/JM over the division ring Λ/J. Show also that $u_1, u_2, \ldots, u_n$ is a minimal set of generators for M if and only if $\bar{u}_1, \bar{u}_2, \ldots, \bar{u}_n$ is a base for M/JM over Λ/J. Deduce that any two minimal sets of generators for M contain the same number of elements and that this number is equal to the length of M/JM considered either as a Λ-module or as a Λ/J-module.*

Solution. If $M = \Lambda u_1 + \Lambda u_2 + \ldots + \Lambda u_n$, then

$$M/JM = \Lambda\bar{u}_1 + \Lambda\bar{u}_2 + \ldots + \Lambda\bar{u}_n.$$

On the other hand, if $M/JM = \Lambda\bar{u}_1 + \Lambda\bar{u}_2 + \ldots + \Lambda\bar{u}_n$, we may conclude that $M = JM + \Lambda u_1 + \Lambda u_2 + \ldots + \Lambda u_n$ and therefore

$$M = \Lambda u_1 + \Lambda u_2 + \ldots + \Lambda u_n$$

by Theorem 4.

These remarks show that the u_i form a minimal set of generators for M if and only if the $\bar{u}_i$ form a minimal set of generators for M/JM. Assume that $\Lambda\bar{u}_1 + \ldots + \Lambda\bar{u}_n = M/JM$ or equivalently that

$$(\Lambda/J)\,\bar{u}_1 + \ldots + (\Lambda/J)\,\bar{u}_n = M/JM.$$

In these circumstances no $\bar{u}_i$ is a superfluous generator if and only if $\bar{u}_1, \bar{u}_2, \ldots, \bar{u}_n$ are linearly independent over Λ/J. Thus the elements $\bar{u}_1, \bar{u}_2, \ldots, \bar{u}_n$ form a minimal set of generators for M/JM when and only when they are a base for M/JM over Λ/J. Assume that this is the case. Then M/JM is isomorphic to a direct sum of n copies of the left Λ-module Λ/J. Accordingly M/JM has length n and all is proved.

Exercise 2. *Λ is a quasi-local ring with radical J and M is a finitely generated left Λ-module. Furthermore $u \in M$ but $u \notin JM$. Show that there exists a minimal set of generators for M that has u as a member.*

Solution. If $v \in M$ we shall use $\bar{v}$ to denote its natural image in M/JM. By Exercise 1, we have only to show that there exists a base for M/JM over the division ring $\Lambda/J = D$ (say) which has u as one of its elements. Put $K = M/JM$ and denote by $\phi : K \to K/D\bar{u}$ the canonical D-homomorphism. There now exist $x_1, x_2, \ldots, x_p$ in K such that

$$\phi(x_1), \phi(x_2), \ldots, \phi(x_p)$$

form a base for $K/D\bar{u}$ over D. Since $\bar{u} \neq 0$, it follows that $u, x_1, x_2, \ldots, x_p$ is a base for K over D.

Exercise 3. *Let I be a two-sided ideal of the arbitrary ring Λ and (P, ψ) a projective cover for the left Λ-module A. Show that $(P/IP, \psi^*)$ is a projective cover for the Λ/I-module A/IA, where $\psi^* : P/IP \to A/IA$ is the homomorphism induced by ψ.*

Solution. By (Chapter 3, Exercise 18), P/IP is a projective Λ/I-module and it is clear that ψ^* is an epimorphism. Let K be a submodule of P satisfying $IP \subseteq K \subseteq P$ and $\psi^*(K/IP) = A/IA$. Then $IA + \psi(K) = A$ and therefore $\psi(IP + K) = A$. We now see that $IP + K = P$ and hence, since $IP \subseteq K$, that $K = P$. This completes the solution.

Exercise 4. *Show that a member of the power series ring $\Lambda[[x_1, x_2, \ldots, x_s]]$ is a unit of that ring if and only if its constant term is a unit of Λ. Show also that if Λ is left Noetherian, then so too is $\Lambda[[x_1, x_2, \ldots, x_s]]$.*

Solution. Let $f \in \Lambda[[x_1, x_2, \ldots, x_s]]$. It is obvious that if f is a unit in the power series ring, then its constant term is a unit in Λ. Now suppose that $f = u + \phi$, where $u \in \Lambda$ and is a unit in that ring, and ϕ is a power series with zero constant term. Then $u^{-1}f = 1 - \omega$, where ω is a power series with zero constant term, and it suffices to show that $1 - \omega$ is a unit. Suppose that $\nu_1 \geqslant 0, \nu_2 \geqslant 0, \ldots, \nu_s \geqslant 0$ are integers. If $k \geqslant \nu_1 + \nu_2 + \ldots + \nu_s$, then the coefficient $a_{\nu_1 \nu_2 \ldots \nu_s}$ of $x_1^{\nu_1} x_2^{\nu_2} \ldots x_s^{\nu_s}$ in $1 + \omega + \omega^2 + \ldots + \omega^k$ is independent of k. Put

$$\psi = \Sigma a_{\nu_1 \nu_2 \ldots \nu_s} x_1^{\nu_1} x_2^{\nu_2} \ldots x_s^{\nu_s}.$$

An easy verification shows that $(1 - \omega)\psi = 1 = \psi(1 - \omega)$. Consequently $1 - \omega$ is a unit as required.

From now on we shall assume that Λ is left Noetherian and we shall show that $\Lambda[[x_1, x_2, \ldots, x_s]]$ is left Noetherian by using induction on s. We begin with the case $s = 1$ and use x to denote the single indeterminate.

Let I be a left ideal of $\Lambda[[x]]$. For $k \geqslant 0$ denote by U_k the set of power series which (a) belong to I, and (b) are such that each of $x^0, x^1, \ldots, x^{k-1}$ has a zero coefficient. Let A_k be the set of elements of Λ formed by the coefficients of x^k arising from the members of U_k. Then A_k is a left ideal of Λ and $A_k \subseteq A_{k+1}$. For each $k \geqslant 0$ we select a *finite* set S_k of members of U_k so that the coefficients of x^k in the various members of S_k generate A_k as a left ideal. Finally we choose an integer m so that $A_k = A_m$ for all $k \geqslant m$.

We claim that $S_1 \cup S_2 \cup \ldots \cup S_m$ generates I as a left $\Lambda[[x]]$-ideal. For let $f \in I$. By subtracting from f a suitable linear combination (with coefficients in Λ) of the members of $S_1 \cup S_2 \cup \ldots \cup S_{m-1}$ we obtain a power series $g \in U_m$.

Let $S_m = \{\phi_1, \phi_2, \ldots, \phi_q\}$. For each i $(1 \leqslant i \leqslant q)$ we can form an infinite sequence $\lambda_0^{(i)}, \lambda_1^{(i)}, \lambda_2^{(i)}, \ldots$ of elements of Λ such that, if $t \geqslant 0$, then

$$g - \sum_{i=1}^{q} (\lambda_0^{(i)} + \lambda_1^{(i)}x + \ldots + \lambda_t^{(i)}x^t) \phi_i$$

belongs to U_{m+t+1}. Then

$$g = \sum_{i=1}^{q} (\lambda_0^{(i)} + \lambda_1^{(i)}x + \lambda_2^{(i)}x^2 + \ldots) \phi_i$$

and our claim follows. Accordingly I is a finitely generated ideal and therefore $\Lambda[[x]]$ is a left Noetherian ring.

Suppose now that $s > 1$ and put $\Lambda^* = \Lambda[[x_1, x_2, \ldots, x_{s-1}]]$. Each member of $\Lambda[[x_1, x_2, \ldots, x_{s-1}, x_s]]$ can be written, in a unique manner, as a power series in x_s with coefficients that belong to Λ^*. Thus the rings $\Lambda[[x_1, x_2, ..., x_s]]$ and $\Lambda^*[[x_s]]$ may be identified and now it is clear how the main conclusion of the last paragraph extends by induction.

Exercise 5. *Let Λ be a left local ring with radical J and C a finitely generated left Λ-module. Show that the following two statements are equivalent*:

(1) *C is Λ-projective*;

(2) *for every exact sequence $0 \to A \to B \to C \to 0$ of Λ-modules the induced sequence $0 \to A/JA \to B/JB \to C/JC \to 0$ is exact.*

Solution. First assume that C is projective, and let $0 \to A \to B \to C \to 0$ be exact. We can define, in an obvious manner, a covariant functor $F: \mathscr{C}_\Lambda^L \to \mathscr{C}_\Lambda^L$ so that $F(U) = U/JU$. Then F is additive and therefore, by (Chapter 1, Theorem 5), it preserves split exact sequences. But $0 \to A \to B \to C \to 0$ splits because C is projective. Accordingly

$$0 \to A/JA \to B/JB \to C/JC \to 0$$

is a split exact sequence and we have shown that (1) implies (2).

Now assume that (2) holds. Let $u_1, u_2, ..., u_n$ be a minimal set of generators for C and let F be a free Λ-module with a base $e_1, e_2, ..., e_n$ of n elements. Define a Λ-epimorphism $\psi: F \to C$ so that $\psi(e_i) = u_i$ and put $K = \operatorname{Ker} \psi$. By (2), the exact sequence $0 \to K \to F \to C \to 0$ induces an exact sequence $0 \to K/JK \to F/JF \to C/JC \to 0$ and, by Exercise 1, both F/JF and C/JC have length n. By (Chapter 5, Theorem 20), K/JK has zero length and therefore $K = JK$. But K is finitely generated because Λ is left Noetherian. Consequently, by Theorem 4, $K = 0$. Thus C and F are isomorphic and, with this, the solution is complete.

Exercise 6. *Let A and B be ideals of S, where B is finitely generated and $B \subseteq \operatorname{Rad} A$. Show that $B^m \subseteq A$ for some positive integer m.*

Solution. Let $B = S\beta_1 + S\beta_2 + ... + S\beta_q$ and choose k so that $\beta_i^k \in A$ for $i = 1, 2, ..., q$. If now $\nu_1, \nu_2, ..., \nu_q$ are non-negative integers such that $\nu_1 + \nu_2 + ... + \nu_q = kq$, then there exists i such that $\nu_i \geqslant k$ and therefore $\beta_1^{\nu_1} \beta_2^{\nu_2} ... \beta_q^{\nu_q} \in A$. Accordingly $B^{kq} \subseteq A$.

Exercise 7. *Let P be a prime ideal of S and let $N_1, N_2, ..., N_q$ $(q \geqslant 1)$ be P-primary submodules of an S-module M. Show that*

$$N_1 \cap N_2 \cap ... \cap N_q$$

is also a P-primary submodule of M.

Solution. Put $N = N_1 \cap N_2 \cap \ldots \cap N_q$. Then

$$\text{Ann}_S(M/N) = \text{Ann}_S(M/N_1) \cap \ldots \cap \text{Ann}_S(M/N_q)$$

and therefore, by (6.5.1),

$$\text{Rad}\{\text{Ann}_S(M/N)\} = \text{Rad}\{\text{Ann}_S(M/N_1)\} \cap \ldots \cap \text{Rad}\{\text{Ann}_S(M/N_q)\}$$
$$= P.$$

Now assume that $s \in S$, $x \in M$, $sx \in N$ and $s \notin P$. Then, since $sx \in N_i$ and s is not in $\text{Rad}\{\text{Ann}_S(M/N_i)\} = P$, it follows that $x \in N_i$. This shows that $x \in N$ and the desired result follows.

Exercise 8. *Let M be a Noetherian Λ-module. Show that every submodule of M can be expressed as a finite intersection of irreducible submodules.*

Solution. Let Σ be the set of submodules of M which cannot be expressed as finite intersections of irreducible submodules. We shall assume that Σ is not empty and derive a contradiction.

Since Λ is Noetherian, we can choose a maximal member K of Σ. Then $K \neq M$ (otherwise K would be an empty intersection of irreducible submodules) nor is it irreducible. Hence $K = K_1 \cap K_2$, where each K_i is a submodule of M strictly containing K. Thus neither K_1 nor K_2 is in Σ and therefore each is a finite intersection of irreducible submodules. But this implies that K itself is the intersection of a finite number of irreducible submodules of M and now we have the desired contradiction.

Exercise 9. *Let Λ be a left Noetherian ring and an S-algebra. Further let M be a finitely generated left Λ-module and A, B ideals of S such that $AM \neq M$ and $BM \neq B$. Show that if $A\Lambda$ contains a power of $B\Lambda$, then $\text{gr}\{B; M\} \leqslant \text{gr}\{A; M\}$.*

Solution. Let $\phi : S \to \Lambda$ be the structural homomorphism and put $C = \phi^{-1}(A\Lambda)$. Then $A\Lambda = C\Lambda$. Consequently $CM \neq M$ and, by (6.6.3), $\text{gr}\{A; M\} = \text{gr}\{C; M\}$. Now choose an integer k so that $B^k\Lambda \subseteq A\Lambda$. This secures that $B^k \subseteq C$.

Let $\gamma_1, \gamma_2, \ldots, \gamma_t$ be a maximal S-sequence on M with

$$\gamma_i \in B \ (1 \leqslant i \leqslant t).$$

Then $t = \text{gr}\{B; M\}$. Further, let us arrange that as many as possible of the terms at the *beginning* of the sequence $\gamma_1, \gamma_2, \ldots, \gamma_t$ are in B^k. *We claim that, in fact, all the γ_i are now in B^k.* For if not, let γ_ν be the first to satisfy $\gamma_\nu \notin B^k$. Then γ_ν is not a zerodivisor on

$$M/(\gamma_1 M + \ldots + \gamma_{\nu-1} M)$$

and hence the same holds for γ_ν^k. Thus $\gamma_1, \ldots, \gamma_{\nu-1}, \gamma_\nu^k$ is an S-sequence on M and, as we know, it can be continued to provide a maximal S-sequence on M in B. However, since $\gamma_\nu^k \in B^k$, we now have a contradiction and with it our claim is established.

It has been shown that all of $\gamma_1, \gamma_2, \ldots, \gamma_t$ are in B^k and we know that $B^k \subseteq C$. Accordingly

$$\mathrm{gr}\,\{C; M\} \geqslant t = \mathrm{gr}\,\{B; M\}$$

and therefore $\mathrm{gr}\,\{A; M\} \geqslant \mathrm{gr}\,\{B; M\}$ as required.

REFERENCES

(1) Auslander, M. and Buchsbaum, D. A. Homological dimension in local rings. *Trans. Amer. Math. Soc.* **85** (1957), 390–405.

(2) Auslander, M. and Buchsbaum, D. A. Homological dimension in Noetherian rings. II. *Trans. Amer. Math. Soc.* **88** (1958), 194–206.

(3) Bass, H. Finitistic homological dimension and a homological generalization of semi-primary rings. *Trans. Amer. Math. Soc.* **95** (1960), 466–88.

(4) Bass, H. Injective dimension in Noetherian rings. *Trans. Amer. Math. Soc.* **102** (1962), 18–29.

(5) Cartan, H. and Eilenberg, S. *Homological Algebra.* Princeton University Press, 1956.

(6) Dieudonné, J. Remarks on quasi-Frobenius rings. *Illinois J. Math.* **2** (1958), 346–54.

(7) Freyd, P. *Abelian Categories.* Harper and Row, 1964.

(8) Hilbert, D. Über der Theorie der algebraischen Formen. *Math. Ann.* **36** (1898), 473–534.

(9) Jans, J. P. Duality in Noetherian rings. *Proc. Amer. Math. Soc.* **12** (1961), 829–35.

(10) Jans, J. P. *Rings and Homology.* Holt, Rinehart and Winston, 1964.

(11) Kaplansky, I. Projective modules. *Ann. of Math.* **68** (1958), 372–7.

(12) Kaplansky, I. On the dimension of modules and algebras. X. A right hereditary ring which is not left hereditary. *Nagoya Math. J.* **13** (1958), 85–8.

(13) Kaplansky, I. *Fields and Rings.* Chicago University Press, 1969.

(14) MacLane, S. *Homology.* Springer, 1963.

(15) Matlis, E. Injective modules over Noetherian rings. *Pacific J. of Math.* **8** (1958), 511–28.

(16) Matlis, E. Applications of duality. *Proc. Amer. Math. Soc.* **10** (1959), 659–62.

(17) Mitchell, B. *Theory of Categories.* Academic Press, 1965.

(18) Morita, K. Category-isomorphisms and endomorphism rings of modules. *Trans. Amer. Math. Soc.* **103** (1962), 451–69.

(19) Northcott, D. G. *An Introduction to Homological Algebra.* Cambridge University Press, 1960.

(20) Northcott, D. G. *Lessons on Rings, Modules and Multiplicities.* Cambridge University Press, 1968.

(21) Northcott, D. G. Generalized R-sequences. *Journ. für reine und angewandte Mathematik*, **239/240** (1970), 7–19.

(22) Rees, D. The grade of an ideal or module. *Proc. Camb. Phil. Soc.* **53** (1957), 28–42.

(23) Rosenberg, A. and Zelinsky, D. Finiteness of the injective hull. *Math. Zeit.* **70** (1959), 372–80.

(24) Serre, J. P. Sur la dimension des anneaux et des modules noethériens. In *Proceedings of the International Symposium on Algebraic Number Theory.* Tokyo and Nikko, 1955; Science Council of Japan, Tokyo, 1956.

(25) Sharpe, D. W. and Vámos, P. *Injective Modules.* Cambridge Tracts in Mathematics No. 62, 1972.

(26) Swan, R. G. *Algebraic K-theory.* Lecture Notes in Mathematics, No. 76, Springer, 1968.

INDEX

The numbers refer to pages